FORENSIC DNA BIOLOGY:
A LABORATORY MANUAL

FORENSIC DNA BIOLOGY: A LABORATORY MANUAL

KELLY M. ELKINS

Assistant Professor of Chemistry, Chemistry Department and Forensic Science Program,
Towson University, Towson, Maryland, United States

AMSTERDAM • BOSTON • HEIDELBERG • LONDON
NEW YORK • OXFORD • PARIS • SAN DIEGO
SAN FRANCISCO • SINGAPORE • SYDNEY • TOKYO
Academic Press is an Imprint of Elsevier

Academic Press is an imprint of Elsevier
The Boulevard, Langford Lane, Kidlington, Oxford, OX5 1GB, UK
225 Wyman Street, Waltham, MA 02451, USA

First published 2013

British Library Cataloguing in Publication Data
A catalogue record for this book is available from the British Library

ISBN: 978-0-12-394585-3

For information on all Academic Press publications
visit our website at **store.elsevier.com**

Printed and bound in the United States

12 13 14 15 10 9 8 7 6 5 4 3 2 1

Working together to grow
libraries in developing countries

www.elsevier.com | www.bookaid.org | www.sabre.org

ELSEVIER   BOOK AID International   Sabre Foundation

Contents

Acknowledgements

I would like to offer a special thanks to my friends and colleagues who provided information, reviewed the manuscript, offered ideas and suggestions, provided photos and provided me with reagents and other materials. These individuals include Susan Berdine, Sandra Bonetti, Carol Crowe, Katie Lobato, Beth Mishalanie, and Francesca Wheeler. I am grateful to my students for their patience while studying forensic biology with me from early drafts of this work. I have appreciated all of your suggestions.

I would also like to thank the four anonymous prospectus reviewers and three manuscript reviewers who provided excellent editing and advice. The professionals at Elsevier including Liz Brown, Acquisitions Editor, Kristi Anderson, Editorial Project Manager, Lisa Jones, Project Manager, and the typesetters and copy editors have been a pleasure to work with and have provided me with helpful assistance and deadlines.

My husband, Tim, served as the initial editor of the manuscript and provided a careful review and helpful suggestions. This made the book more coherent and readable. I am indebted to the support and patience of my children, Madeleine and Katie, who have given up many hours with me so that I could finish this project. Thank you to my sister, Melanie, and my brother-in-law, Will, for your kind support and many hours of free babysitting. Thanks also to my parents who have never given up believing in me.

I was first exposed to forensic DNA typing in only 2005 as a Temporary Assistant Professor at Armstrong Atlantic State University. Thank you for the opportunity you gave me in allowing me to teach Chemical Forensics to your students. Since then, I have also had the opportunity to teach forensic courses at Keene State College and Metropolitan State University (formerly College) of Denver (Metro State). However, I am indebted to Chris Tindall for taking a chance on me to lauch Metro State's new Criminalistics II course in Spring 2008. Without his guidance, ideas, suggestions and support, this project would not have been possible.

Manuscript Reviewers

Tim Frasier
St. Mary's University

Sarah Adamowicz
University of Guelph

Margaret Wallace
John Jay College

About the Author

Dr. Kelly M. Elkins is an Assistant Professor of Chemistry at Towson University. She earned her B.S. degree in Biology and B.A. degree in Chemistry from Keene State College in 1997 and her M.A. and Ph.D. degrees in Chemistry from Clark University in 2001 and 2003, respectively. She was a Fulbright Scholar in Heidelberg, Germany from 2001-2002 and a Cancer Research Institute postdoctoral fellow at MIT from 2003-2004. Previously, she held positions as Assistant Professor of Chemistry and Director of Forensic Science at the Metropolitan State University of Denver, Temporary Assistant Professor of Chemistry at Armstrong Atlantic State University, and Adjunct Professor of Chemistry and Biology at Keene State College, Cloud County Community College and Highland Community College.

She has had an active research group for the past several years that has focused on DNA recovery and methods development, applying instrumental tools to the detection of body fluids, DNA cloning, and nanoparticle synthesis and applications. She has supervised more than twenty undergraduate and high school student research projects, authored thirteen scientific publications, and has delivered more than 50 conference and seminar presentations. She has served as a peer reviewer eight referred journals and four textbook publishers. Her research has been funded by a Petroleum Research Fund Summer Research Fellowship, ACS Project SEED, and the Forensic Sciences Foundation. She is a member of the American Chemical Society (ACS) and served as an elected Alternate Councilor of the Colorado section, Executive Committee and as Co-Chair of the Student Grants Committee. She is an Associate Member of the American Academy of Forensic Sciences and President of the Council of Forensic Science Educators (2012). She has appeared as a forensic expert on Denver television stations ABC 7 News, Fox 31 News, and NBC 9News and has been interviewed by *The Denver Post* and *Forensic Magazine*.

Welcome

Forensic DNA Biology: A Laboratory Manual details more than 20 step-by-step laboratory experiments in deoxyribonucleic acid (DNA) typing, covering such as topics as short-tandem repeats, paternity testing, single-nucleotide polymorphisms, DNA cloning, statistical analysis, and social, ethical, and regulatory concerns. The manual is designed to provide upper-level undergraduates and others who are new to the field a fundamental understanding of the practical application of forensic DNA analysis, including evidence collection best practices, DNA extraction techniques, quantitation, and typing analysis. The labs contained herein are designed to meet the laboratory needs of a forensic DNA biology course by offering substantial background information to marry theory with practice.

The laboratory manual includes an introduction to the field of forensic DNA biology and laboratory safety and examples and exercises for performing statistical calculations. The experiments cover the identification, collection, packaging, and handling of biological evidence; the proper use of micropipettes; serological testing; DNA extraction; agarose gel electrophoresis; quantitation using ultraviolet visible (UV-vis) spectroscopy, fluorescence spectroscopy, and real-time polymerase chain reaction (PCR); multiplex PCR primer design and PCR; cloning; multiplex PCR amplification and capillary electrophoresis of short-tandem repeat (STR) and Y-chromosome STR loci using commercial kits; single nucleotide polymorphism (SNP) detection of mitochondrial DNA using real-time PCR; visualization and analysis of DNA sequence data; RNA extraction and body fluids analysis; and DNA extraction typing of botanical material.

This manual adapts laboratory experiments and procedures commonly found in the literature and employed by regional, state, and federal crime laboratories. In addition, these experiments are designed to be executed in a standard academic laboratory with standard molecular biology equipment; indeed, where possible, this manual employs and recommends the use of inexpensive versions of common procedures.

Forensic DNA Biology: An Introduction

Forensic science is a broad field that includes the disciplines of pathology, psychiatry, engineering, computer science, toxicology, odontology, anthropology, and botany, as well as the chemical, physical, and biological sciences. Criminalistics is a branch of forensic science and centers on the application of the principles of the aforementioned field including the chemical, physical, and biological sciences to civil and criminal law. Criminalistics focuses on the recognition, documentation, collection, preservation, and analysis of physical evidence. A criminalist is a specialist who uses scientific principles to analyze, compare, or identify firearms (ballistic evidence), fingerprints, hairs, fibers, drugs, blood, and other physical evidence. The goal of criminalistics is to positively identify the source of physical evidence in order to provide law enforcement officials with a connection to a crime scene (Locard's Principle of Transference).

One of the most effective ways to accomplish this goal is through the examination and analysis of biological evidence, notably material that contains deoxyribonucleic acid (DNA). Early criminalists used chemical assays, or investigative procedures, to determine the presence of biological materials in stains as a means of establishing the likelihood that a given piece of evidence originated with a specific individual. Later, criminalists experimented with antigen polymorphism, or blood group typing, to identify the presence or absence of inherited antigenic substances on the surface of red blood cells (RBCs). When linked to other forms of evidence, this method sought to scientifically link an individual or individuals with a particular crime by linking them to the victim or the crime scene.

More recently, criminalists have turned to DNA typing, or DNA profiling, as a means of scientifically identifying the originator of biological crime scene evidence. Indeed, DNA typing has revolutionized criminal investigations and has become a powerful tool used to identify individuals in criminal, paternity, and missing persons cases. And although it is possible to exclude an individual as a source of a biological sample using antigen-antibody interactions, protein, or enzyme polymorphisms, only DNA typing provides sufficient discriminating power to positively identify beyond a reasonable doubt the originator of biological criminal evidence.

BIOLOGY OVERVIEW

DNA may be extracted from almost any biological material, including hair cells and body fluids (except red blood cells). DNA is present in every cell in the nucleus and in the mitochondria of animals. Plants also contain chloroplasts, which contain additional DNA in the form of a circular chromosome. DNA the instructions used to transmit, encode, and express genetic information. DNA and a related molecule, ribonucleic acid (RNA), belong to a class of compounds called nucleic acids. Nucleic acids are polymers composed of monomer units called nucleotides, each with one of a variable nitrogenous base, adenine, thymine, guanine, and cytosine, and known by the letters A, T, G, and C, a phosphate group and a five-carbon sugar. Of the nitrogenous bases, thymine is found only in DNA whereas uracil is found in RNA. The DNA segments that carry genetic information are called genes. DNA genes are chemical substances composed of a specific sequence of nucleotides; these genes code for proteins or other gene products.

In 1953, James Watson and Francis Crick proposed a three-dimensional structure for DNA to explain its chemical and physical properties. Their model of DNA consisted of two helical polynucleotide chains coiled around an axis and connected through hydrogen bonding, forming a double-helix. The outside of the helix contains hydrophilic sugars and negatively charged phosphate groups of the individual nucleotides; the inside of the helix contains the hydrophobic bases. The nucleotides making up each strand of DNA are connected by phosphodiester bonds between the phosphate group and the deoxyribose sugar on the outside of the helix. This forms the backbone of the DNA strand from which the nitrogen-containing bases extend into the center of the helix. The bases of one strand of DNA will pair in a complementary fashion with nitrogen-containing bases on the other strand through Watson-Crick hydrogen bonding.

Each strand of the DNA double-helix is not identical, but rather complementary. That is, the structure of adenine permits it to form two hydrogen bonds only with thymine when a part of the double-helix, and cytosine will form three hydrogen bonds only with guanine. In other words, where thymine appears on one strand, adenine will appear on the other. This makes DNA an informational biomolecule in which the sequence of one strand of the DNA helix can be predicted exactly from only the sequence of the complementary strand.

DNA profiling is simply the collection, processing, and analysis of the unique sequences (AGTGAAGTCGAAC), base polymorphisms (AGTGA**A**GTCGAAC vs. AGTGA**T**GTCGAAC), or tandem repeats (for example, AATGAATGAATG) of A, T, G, or C nucleotides found on the chromosome of humans and other organisms. The human genome comprises approximately 3.2 billion base pairs. The DNA is coiled and packaged for storage in each cell around histone proteins and supercoiled in the form of chromosomes. Humans have 46 chromosomes, including pairs of 22 autosomes (numbered from 1 to 22) and a pair of sex chromosomes in all cells except gametes, egg and sperm sex cells which have one copy of the 22 autosomal chromosomes and one sex chromosome. Cells containing all 46 chromosomes are referred to as diploid, whereas sex cells are haploid. Each biological parent provides the DNA in one of the 22 autosomes and one sex chromosome. Females are characterized as having received two X sex chromosomes, one from each parent. Males are characterized as having received an X and a Y sex chromosome. The biological mother always gives an X chromosome, so the biological father determines the sex of the offspring.

Corresponding regions of the same numbered pair of chromosomes, homologous chromosomes, are termed *loci*. Alleles are variations of nucleic acid base pairs or sequences at a given locus, and they may be alternate forms of genes including dominant and recessive traits. Individuals with identical bases or sequences of bases at a locus on their homologous chromosomes are termed *homozygous* for alleles at that locus. Individuals with different bases or sequences of bases at a locus on their homologous chromosomes are termed *heterozygous* for alleles at that locus. In other words, the chromosomes inherited from the biological parents may contain identical or nonidentical sequences or bases at a locus, and the offspring may have two copies of the same allele or one copy of two different alleles at each locus. In different experiments, DNA typing may involve the evaluation of the similarities or differences in base pairs at loci between individuals, exact base sequences of pieces of chromosomes, or base polymorphisms such as numbers of repeating sequences of DNA.

RESTRICTION FRAGMENT LENGTH POLYMORPHISMS

Forensic DNA typing debuted with the evaluation of restriction fragment length polymorphisms (RFLPs). The RFLP typing method involved evaluating the presence and number of a variable number of tandem repeats (VNTRs), minisatellites that are approximately 10- to 60-nucleotide base pair sequences that are repeated in the genome approximately 1000 times. Directly adjacent VNTR repeats were detected by digesting, or cutting, the DNA with restriction enzymes or, later, by amplifying the DNA using the polymerase chain reaction (PCR), and evaluating the size of the product using gel electrophoresis. This technique, which represented the first method of multilocus DNA typing, was developed by Alec Jeffreys in Leicestershire, England, and first published in 1985 in the journal *Nature*. The forensic science community recognized the applications of this technique, and Dr. Jeffreys was soon receiving requests by legal authorities to use

this approach to evaluate evidence from crime scenes. One early example helped law enforcement officials exonerate a man who had falsely confessed to the rape and murder of two young girls in 1983 and 1986. Analysis of semen stains from the crime scene in 1987 subsequently led to the conviction of another suspect in 1988.

Almost overnight, DNA typing changed the investigation and prosecution of criminal cases. In addition, this technique opened possibilities of solving cold cases and other hard-to-solve cases where eyewitness accounts were nonexistent or inconclusive or where the quantity and quality of physical evidence prevented the identification of a perpetrator.

RFLP typing did have its drawbacks, however. This technique required a large amount of high-molecular-weight, double-stranded DNA. However, many samples from crime scenes provided only small amounts of DNA, and much of this material degraded into smaller fragments depending on the conditions of the crime scene or storage techniques. In addition, the RFLP analysis process took a long time, approximately 6 to 8 weeks to process evidence. Another drawback was the need for radiation probes for visualization. Finally, RFLPs were sorted into "bins," or ranges of sizes of DNA fragments, rather than exact sizes by numbers of base pairs. Thus, although the method had tremendous power to discriminate and eliminate individuals, the drawbacks were enough to encourage the scientific community to continue to search for better DNA typing methods.

POLYMERASE CHAIN REACTION

In 1983, Kary Mullis invented the polymerase chain reaction (PCR) technique now widely employed for forensic DNA typing. With PCR, it became possible for DNA analysts to examine variation in genetic characteristics using only a small amount of DNA, theoretically using only a single cell (or only 6 picograms of DNA in a diploid cell in humans). In 1986, the multiallelic human leukocyte antigen HLA-DQα locus was presented for forensic DNA typing. Initially, it was evaluated using RFLPs, but PCR-based typing was also later applied to this locus. VNTR minisatellites were not particularly amenable to PCR because of the long base pair sequences. By typing VNTR microsatellites, DNA samples did not have to be in long fragments but could be identified in shorter fragments of two- to seven-base pairs, known as short tandem repeats (STRs) or simple sequence repeats (SSRs).

SHORT TANDEM REPEATS

STR typing dramatically reduced the time and cost of DNA analysis, from 6 to 8 weeks to approximately 1 to 2 days. And at $400 per sample, this methodology is less expensive than many alternative methods. PCR-based methods use fluorescence detection, thereby eliminating the need for radioactive probes for visualization. Further, this method could evaluate single-stranded and degraded DNA as well as simultaneously evaluate multiple loci in a single test tube. High-quality ladders and internal standards allowed analysts to match the fragment size to determine the number of repeats at a given humidity and temperature. STR loci occur approximately once per every 15,000 bases in a genome and have a mutation rate of 1 in 1000. The selected loci are mostly distributed on different chromosomes. At a given locus, the number of STR repeats is referred to as an allele. By

evaluating the number of tandem repeats at each of 13 STR loci, scientists have determined that they can also discriminate genotypes to one in trillions using this method. In addition to requiring very small quantities of DNA, mixture components can be differentiated, as the markers can be highly polymorphic as in having many possible alleles and genotypes in a population.

In addition, STRs and other mostly bi-allelic loci located on the X and Y sex chromosomes can be employed in DNA typing, including sex typing. Polymorphic X- and Y- chromosomal STRs have also been employed in paternity and maternity testing, familial typing, and mixture analysis in sexual assault cases. As there is only one copy of the Y chromosome occurring only in males, the interpretation is simpler than with autosomal STRs. Y typing is highly effective for discerning DNA mixtures of male and female DNA and for tracing ancestral lineages and biological paternity in male lines.

SINGLE NUCLEOTIDE POLYMORPHISMS

Another type of molecular marker that is gaining popularity in the forensic community is single nucleotide polymorphisms (SNPs). The analysis of SNPs involves identifying specific positions within the genome where the A, C, G, or T base is known to vary across individuals. By analyzing samples using a large panel of these variable sites, forensic scientists can identify individual-specific bi-allelic genotypes by focusing on single bases differences that can be used for exclusion and individualization of a given sample of material.

One advantage of SNPs over STRs is that they require extremely small fragments of DNA, and therefore SNP analysis can be successful with samples that are too degraded for STR analysis. Another advantage of SNPs is that they are much more common throughout the genome than STRs, with one SNP occurring approximately every 1000 bases in the genome, and they have a low mutation rate of 1 in 1,000,000,000. One drawback of SNPs is that they have much less variability than STRs. With SNPs, only one of four alleles is possible (A, C, G, or T), whereas some STR alleles have more than 20 alleles. As a result, many more SNP markers (estimated to be 50 or more) are required to achieve the same level of resolution as a much smaller number of STRs.

Widespread use of SNP technology has not spread as quickly as that of STRs. There are already more than 9 million STR entries into the U.S. National Combined DNA Index (CODIS) database in 2011 and millions more entries in other databases worldwide (e.g., more than 5 million in the United Kingdom's NDNAD database alone). Adoption of a new methodology necessitates inclusion of SNP data in databases that will be useful for searching for individuals, including convicted offenders and missing persons, and computing rarity statistics. Nonetheless, SNPs have proved valuable and are employed in estimating ethnic background, predicting some phenotypes and physical traits, and evaluating severely degraded DNA samples.

MITOCHONDRIAL DNA

Since 1996, laboratories have also employed mitochondrial DNA (mtDNA) typing. Mitochondrial DNA typing has found uses in the evaluation of degraded DNA samples and in familial typing in mass disaster situations. Mitochondrial DNA is transmitted maternally via the egg cell. Copies of the 16,569 base pair mitochondrial chromosome can exceed a thousand in a single cell. The mutation rate is five to ten times higher than that of nuclear DNA. Three highly variable regions, HVI, HVII, and HVIII, are differentiated by DNA sequencing or by a "hotspot" analysis in which SNPs alleles are identified within these and other regions. The discrimination power is limited to the size of current databases, or one in thousands for mtDNA. The required time for analysis ranges from a few days to 2 weeks depending, in part, on how many times a sequence or set of SNPs needs to be verified.

KNOWN VERSUS QUESTIONED SAMPLES

All of these methods rely on having another sample (known or K) to compare to the evidence (questioned or Q) sample. These typically include reference samples from the suspect or victim in criminal cases or biological family members in paternity, mass disaster, or missing persons cases. The sample may be compared to samples already in a database. Elimination samples from persons known to come in contact with the evidence, including laboratory staff and police staff, are used to ensure that the sample was not a match to them and can be associated with the crime scene. Standard samples of DNA derived from cell lines in which the exact DNA profile is known are used to validate that the reagents and equipment are working properly. Based on the comparison, the DNA analyst will assign a conclusion of an inclusion (also known as a match, failure to exclude, or is consistent with), exclusion (or nonconsistent with, non-match), or inconclusive in the case of lack of sufficient data to make a determination or conflicting data in a profile (e.g., a partial profile match or one locus that does not match in a full profile). The process of forensic DNA typing is diagramed in Figure 0.1.

The number of unique sequences in a given chromosome varies between unrelated individuals, enough that unrelated individuals are highly unlikely to have the same set of nucleotide repeats or DNA base sequences. Thus, by determining the probability that a given sequence(s) or number of repeats of nucleotides belongs to a given individual, criminalists are able to scientifically match the evidence to the source of the evidence and assign a probability of a random match to an unrelated individual in the population based on a population database or a match to a reference or elimination sample.

Scientists from public and private labs are involved in evaluating evidence, reference, and elimination data and may provide their findings and interpretation to the district attorney, defense attorney, family members, judge, jury, and other concerned parties.

WHY STUDY FORENSIC DNA BIOLOGY?

There are significant opportunities in this field, both in research and in the opportunity to impact society through effective law enforcement. A recent search in the *Journal of Forensic Sciences* returned 534 hits for DNA typing and 644 hits for DNA (January 17, 2012). In 2007, Forensic Science International launched a new journal *Forensic Science International: Genetics*. A search returned 467 hits for DNA typing and 569 hits for DNA. This amounts to almost a paper a day published on this topic in 5 years in these two journals alone! The papers range from population studies to new methodology, new loci, new extraction and separation methods, and statistical analysis. The DNA typing methods cited range from

FIGURE 0.1   Process of forensic DNA typing.

TABLE 0.1   Relevant Scientific Working Groups

Group	Abbreviation	Website
Scientific Working Group for DNA Analysis Methods	SWGDAM	None
Scientific Working Group on Disaster Victim Identification	SWGDVI	www.swgdvi.org
Scientific Working Group on Bloodstain Pattern Analysis	SWGSTAIN	www.swgstain.org
Scientific Working Group on Microbial Genetics and Forensics	SWGMGF	None
Scientific Working Group on Chemical, Biological, Radioactive and Nuclear Weapons	SWGCBRN	None
Scientific Working Group on Wildlife Forensics	SWGWILD	None

RFLP to STR and SNP and mitochondrial DNA analyses. DNA profile matching is based on statistical analyses. These have included averages, percentages, standard deviations, and likelihood ratios of data using Mendelian genetics, Bayesian statistics, and combined probability of exclusion to determine the rarity of a random match.

For more than 20 years now, the U.S. Federal Bureau of Investigation (FBI) and several other federal agencies have supported the efforts of various scientific and technical working groups (SWGs and TWGs, respectively) for the advancement of forensic standards and techniques. SWGs/TWGs consist of representatives from the fields of forensic, industrial, commercial, academic, and in some cases international communities. These multidisciplinary working groups assist in developing standards and guidelines and improve communications throughout their respective disciplines (Table 0.1).

Laboratory Safety

RULES FOR A SAFE LAB ENVIRONMENT

Laboratory safety involves a cautious attitude, respect for the materials employed, and an awareness of the potential hazards that can come from the improper use or handling of those materials. Adherence to safety precautions can help prevent most accidents from occurring in the first place. Special note should be taken of specific instructions that are given in an experiment to eliminate recognized potential hazards.

A. General Regulations

1. No visitors are permitted in the laboratory without prior approval (e.g., instructor, lab manager, etc.).
2. No food or beverages are permitted in the lab at any time. No gum or tobacco products are permitted in the lab. Do not taste the chemicals in the lab.
3. Everyone in the laboratory, including visitors, must wear eye protection such as safety glasses or goggles.
4. Clothing should cover all areas of the body. Shorts and tank tops are not allowed.
5. Hair should be pulled back.
6. Shoes should provide full coverage and be made of a chemical-resistant material. No heels or open-toed shoes/sandals will be allowed in the lab.
7. Use of lab coats is recommended.
8. Perform only authorized experiments/techniques. Refer to the appropriate Material Safety Data Sheets (MSDS) for chemical safety.
9. Keep lab benches clean and free of unnecessary items.
10. Do not pipette by mouth.
11. Aliquot from stock solutions; do not pipette directly.
12. Use your hand to waft the odor to your nose if you are directed to note an odor in an experimental procedure.
13. Know the location and usage of safety equipment including first-aid kit, fire extinguisher, eyewash, safety shower, fire blanket, exits, and chemical and biological safety hoods.
14. Read the appropriate procedure prior to entering the lab.
15. When possible, do not work alone. Always have a partner when you are working in the lab.
16. Clearly label all chemicals with your initials, chemical name, and date. Read all labels carefully before dispensing.
17. Examine glassware for cracks and chips prior to use. Promptly report breakage of glassware to your instructor. Put broken glass into its own marked waste container.
18. Wash your hands with soap and water prior to exiting the lab.
19. Do not allow burners or ignition sources when working with flammable liquids.
20. Use care to not puncture your skin when using needles, razor blades, or other sharps.
21. Keep chemicals away from cuts, bruises, or any other places your skin is broken.

22. Use laboratory hoods whenever using flammable and hazardous materials or when the evolution of flammable and hazardous gases is possible.
23. Use of personal electronic devices including cell phones, web devices, and computer gaming and social networking systems is not permitted in lab.
24. Failure to comply with laboratory rules and regulations will result in expulsion from the laboratory.

B. Specific Regulations

25. Gloves should be worn as directed by your instructor or MSDS documents to include when handling strong acids and bases as well as mutagenic and carcinogenic agents to avoid exposure to these chemicals and to protect samples from contamination from nucleases on the hands.
26. If reactions are left overnight, provide your name and your instructor's name and contact information in case of fires, spills, or explosions. Do not leave power equipment such as power stirrers, shakers, heating blocks, hot plates, thermocyclers, and capillary electrophoresis instruments on without the instructor's consent. Check unattended reactions periodically.
27. Dilute aqueous waste between pH 5 to 8, sugar, syrup, soap, and sodium chloride can be disposed of in the sink with water. Dispose of all other wastes (e.g., organic, heavy metals) in the appropriate specifically labeled waste container as directed by your instructor or lab manager.
28. Paper or solid waste can be disposed of in the trash can unless otherwise directed by your instructor.
29. Secure benchtop centrifuges so that they do not "walk" off the bench if vibration

occurs. If vibration occurs, immediately stop the centrifuge and check that an appropriate counterbalance is in place. Always close and keep closed the centrifuge lid while running. Wait until the centrifuge is at full speed and appears to be running safely before leaving the area. Do not use your hand to stop the centrifuges.
30. Clean centrifuges, thermocyclers and other laboratory equipment regularly as directed by your instructor to avoid contamination issues.
31. Use care when working with electrical and voltage equipment. Turn off all sources and disconnect prior to checking gels.
32. Operate UV radiation systems in a completely closed UV box. Always wear UV-absorbing lenses while operating UV light sources. Avoid skin exposure; wear long pants and long-sleeved shirts to protect skin. Do not direct UV light sources at people.
33. Each individual is to report any accident or spill, no matter how small, to the laboratory instructor. If necessary, the laboratory instructor will give a written report of the incident to the department chair or lab director.

C. Contract for Students in Academic Labs

If you do not submit the signed contract to your laboratory instructor, you will not be allowed into the laboratory or be allowed to perform any laboratory work.

I, the undersigned, have read the discussion of good laboratory safety rules and practices presented in this laboratory manual. I recognize it is my responsibility to observe these practices and precautions while present in the laboratory. I understand if I do not comply with these regulations, my instructor will ask me to leave the laboratory and I will receive a grade of zero for that experiment.

_____

Student

_____

Date

Reference

"Safety in Academic Laboratories," American Chemical Society, Seventh ed., 2003.

Avoiding Contamination Issues: Standard Laboratory Practices

1. Clean, dedicated lab coats are worn when working with samples and preparing solutions.
2. Gloves are worn when working with samples. Wash gloved hands with ethanol prior to working with samples. Gloves are removed when leaving the work area/lab or after working with a sample to avoid the transfer of DNA into other work areas or samples.
3. Gloves are changed whenever they may have become contaminated with DNA in order to reduce the unnecessary contamination of the work area/lab with that DNA.
4. The lab bench should be washed with a 10% bleach solution followed by a 95% ethanol/reagent alcohol rinse. Clean paper or absorbent pads may be placed on bench tops when the area is in use, as desired, and are changed as necessary. If a lab bench is contaminated, the absorbent pads should be folded inward and placed into an autoclave bag prior to disposal according to standard lab procedures.
5. Removal of DNA from equipment and supplies is accomplished by either UV light exposure for a minimum of 20 minutes per surface area or by using a 10% bleach solution.
6. Biological waste should be double-bagged in autoclave bags and taken directly to the autoclave.
7. All samples should be clearly labeled.
8. Sterilized microcentrifuge tubes and sterile aerosol-resistant pipette tips are used.
9. Aerosol-resistant and other pipette tips are changed between each sample. The same tip may be used to aliquot kit reagents, buffers, or other liquids that do not contain DNA.
10. Only one DNA-containing tube is open at a time.
11. Tubes containing DNA and other reagents in liquid are centrifuged before opening to reduce aerosol-dispersion and sample loss upon opening.
12. Equipment (centrifuges, pipettors, racks, thermocyclers, etc.) is cleaned as needed using 10% bleach or 20 minutes of UV light exposure.
13. Amplification should be performed in a different location from sampling the evidence/simulated evidence samples and extraction to minimize contamination. If a thermal cycler becomes contaminated, all of the tubes should be removed and discarded. Completely clean the block and affected surfaces with a 10% bleach solution.
14. Controls (extraction and amplification) and blanks (reagent and negative amplification) are run along with samples to check for potential contamination. Reference (known samples from victim, suspect, or family members) and elimination (crime scene investigators, DNA personnel) samples

are used to compare to samples that exhibit the possibility of a mixture.

15. Lot numbers of reagents are recorded in case of failure or suspected contamination.

16. DNA and reagents are stored in the $-20°C$ or $-80°C$ freezer after use. Chemicals are stored as directed by instructor.

Reference

Merritt, K., Hitchins, V.M., Brown, S.A., 2000. Safety and cleaning of medical materials and devices, Journal of Biomedical Materials Research (Applied Biomaterials) 53, 131—136.

1

Pipetting

OBJECTIVE

To learn how to use calibrated variable volume micropipettes and graph the results using Microsoft Excel.

SAFETY

Wear goggles in the lab. Report and clean up all spills immediately. Carefully dry spills on the balances. Keep micropipettes vertical and handle with care.

MATERIALS

1. Variable volume micropipettes (to include some or all of the following: 0.1 to 2.5 μL, 0.1 to 10 μL, 1 to 20 μL, 1 to 100 μL, 10 to 1000 μL, 500 to 5000 μL) and pipette tips referred to as P2.5, P10, P20, P100, P1000, and P5000, respectively
2. Beakers or weigh boats
3. Balance
4. Deionized water
5. Ethanol

BACKGROUND

Calibrated and variable volume adjustable micropipettes are indispensable tools for the forensic scientist. Accurate results will only come from the use of properly calibrated, precise tools. Variable volume micropipettes deliver precise and accurate volumes of liquid, are easy to use, and are available in a variety of sizes. This laboratory will introduce you the skills required for the accurate use of pipettes. In addition, this lab will introduce you to the use of graphing software to analyze the data you collect and prepare reports based on your findings. This lab is a chance to test the accuracy of your equipment.

Forensic DNA Biology

   Forensic scientists also use automatic pipettes to automatically and accurately transfer small liquid volumes between storage containers such as test tubes and platforms for analysis. Each pipette is equipped with a volume indicator, shaft to load the pipette tip, volume adjustment knob, tip ejector button for removing the tip, and a three-position plunger used to load and expunge the liquid. Figures 1.1 to 1.3 show various pipettes, identify the volume range, and depict specially designed pipette tips that transfer specific, nonadjustable volumes of material.

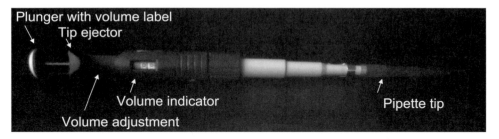

FIGURE 1.1   A fully assembled continuously adjustable rotary pipette.

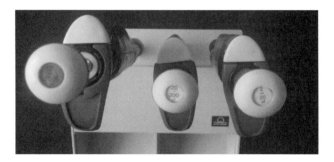

FIGURE 1.2   Pipettes of variable volume ranges (left to right: 100 to 1000 μL, 10 to 200 μL, and 2 to 20 μL) allow the transfer of a specified volume within the pipettes' volume range.

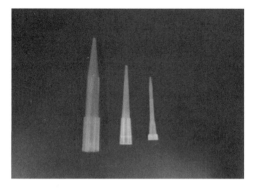

FIGURE 1.3   Specially designed pipette tips (left to right: 100 to 1000 μL, 10 to 200 μL, and 0.5 to 10 μL) that transfer specific, nonadjustable volumes of material.

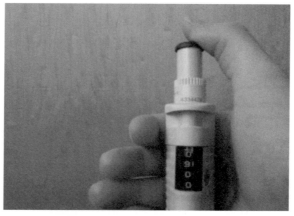

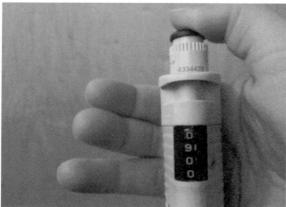

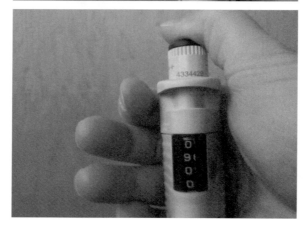

FIGURE 1.4 Use of the depressible plunger to transfer liquids.

Micropipettes are piston-driven air-displacement instruments that contain a plunger to provide suction to pull the liquid in which the pipette tip is immersed up into the disposable tip when the piston is compressed and released using the plunger. For adjustable volume pipettes, the delivery volume is determined by the setting on the volume indicator using a dial. The depression of the plunger causes the piston to move upward and generates a vacuum in the airtight shaft left vacant by the piston. The air in the pipette tip rises into the shaft to fill the space left vacant, and the liquid drawn into the tip fills the space vacated by the air (Martin, 2001).

Table 1.1 shows the maximum volume for some pipettes and how a pipette can be set appropriately to transfer the desired volume in microliters.

TABLE 1.1   Pipette Settings to Deliver the Desired Volume

Pipette (Eppendorf)	Maximum Volume (µL)	Maximum Pipette Setting	Sample Pipette Setting	Sample Pipette Volume (µl)
P1000	1000	1000	900	900
P100	100	1000	900	90.0
P2.5	2.5	2500	900	0.900

The pipettes depicted here require the use of a disposable tip. To use the pipette, do the following:

1. Load a tip from a tip box.
2. Adjust the volume to be transferred using the volume adjustment rotating dial.
3. Hold the pipette vertically, and depress the plunger to the first stop point.
4. Immerse the tip into the liquid material being sampled.
5. While keeping the tip immersed, slowly release plunger to draw up the liquid, pausing momentarily at the end for viscous samples.
6. Insert the tip of the pipette into to the tube or container where the liquid will be deposited.
7. Depress the plunger fully to the second stop point to deposit the liquid into the desired tube.
8. Remove the used tip using the ejector button into an appropriate refuse container.

Note: *Do not* set the pipette to a volume that exceeds the labeled volume or the pipettes will quickly go out of calibration. *Do not* put the pipette barrel directly into the liquids.

In this lab, you will pipette water into a small beaker, or weigh boat. You will record the accuracy of your pipetting by recording the mass of the material you deposit on a balance.

Your precision will be evaluated by replicating a desired volume 10 times. At 25° C, 1 mL of pure water weighs 0.997044 g, with a probability error of $+/- 0.001$ g or $+/- 0.0001$ g because of the balance. Delivering 100 µL of water using a P100 pipette should weigh 0.100 g.

TABLE 1.2   Density of Water at Room Temperatures (CRC 62nd edition)

Temperature (°C)	Water Density (g/mL)
20	0.998203
21	0.997992
22	0.997770
23	0.997538
24	0.997296
25	0.997044

Finally, you will use a spreadsheet and graphing software such as Microsoft Excel to analyze and graph the data you collect. You will summarize your findings in a written lab report that describes your accuracy and precision.

Accuracy is how close a measured value is to the actual (true) value (even after averaging). Precision is how close the measured values are to each other (Figure 1.5). Note that even inaccurate pipettes can have high precision (reproducibility).

Low Accuracy   High Accuracy   High Accuracy
High Precision   High Precision   Low Precision

FIGURE 1.5   Examples of precision and accuracy.

Reminder: Numbers (measurements) should *always* be accompanied by units with the appropriate significant figures!

PROCEDURE

Part A: Using a Pipette

1. Obtain some deionized water with a beaker. Pipet the maximum volume using a selected pipette (e.g., exactly 1 mL using a P1000 pipette).
2. Obtain a weigh boat and place it on the balance and press tare. Pipet the water into the weigh boat. Record the mass. Is it 0.997 g (at 25° C)? If not, adjust your technique until you are close to the expected value.
3. Pipet an intermediate volume (e.g., 557 μL using a P1000 pipette) into the weigh boat. Record the mass. Discard the volume into a beaker.

Part B: Evaluating Precision for Three Pipettes

1. Set a pipette to the maximum volume (e.g., 1000 μL of water 10 times using a P1000). Deliver that volume and record the mass. Repeat nine times for a total of 10 replicates.
2. Repeat step 1, using two more pipettes (e.g., P2.5, P20, P100, P5000) of your choice.

Part C: Evaluating Nominal versus Actual Volume Using a Pipette

Using a pipette of your choice, dispense five equally spaced volumes of water. For example, a P1000 pipette may be used to dispense volumes of 0.2, 0.4, 0.6, 0.8, and 1 mL. Record the volumes set and the mass of the volume delivered for each point (Table 1.3). Repeat two more times. Repeat previous steps using ethanol (density 0.79 g/mL at 20° C), also performing three replicates.

QUESTIONS

1. Did the pipette deliver the mass/volume of water expected for the temperature of the room (accuracy)? If you needed to adjust your technique, did this improve the delivery?
2. Tabulate your data for the three pipettes used to deliver the maximum volumes 10 times. Compute the average mass of the water delivered 10 times and the standard deviation. Show your work. Which of the pipettes was most precise?
3. Compare and contrast pipetting water and ethanol liquids.
4. Using a graphing program, plot the "actual volume dispensed" versus the "pipette setting/ nominal volume" as shown in the sample data (Figure 1.6) for water and ethanol. Calculate the "best" fit line (trendline) for this data using linear regression, and add the trendline, line equation, and R^2 value to the graph. Be sure to label your graph with a title and x-axis and y-axis labels with units. Add error bars (if multiple replicates were performed).

GRAPHING THE DATA USING MICROSOFT EXCEL (2003)

Type your data into two columns. The first column will contain the x-axis; the second column will contain the data plotted on the y-axis. Highlight all of the data by depressing the left button on your mouse. Select Insert, Chart, XY Scatter, next, and finish. Click anywhere on the chart; select Chart from the main menu again, and add trendline, linear, click on options, then select in the boxes R^2 value and line equation on the chart. You can alternately double-click with the left mouse button on any aspect of the chart to alter the presentation format, including axis labels.

EQUATIONS

Average

$$x_{avg} = \sum (a_1 + a_2 + a_3 + \ldots + a_n)/n$$

where a is a delivered mass for a replicate, and n is the number of replicates.

Standard Deviation

$$S_N = \sqrt{1/N\text{-}1} \sum (x_i + x_{avg})^2$$

where x_{avg} is the average (computed previously), x_i is a delivered mass for a replicate, and N is the number of replicates.

TABLE 1.3   Sample Data Table for Part C

Actual Mass (g)	Actual Volume (mL)	Pipette Setting (mL)
0.1814	0.1819	0.200
0.4004	0.4016	0.400
0.5897	0.5915	0.600
0.7998	0.8022	0.800
1.0030	1.0060	1.000

* The density of water is 0.997044 g/mL at 25.0 °C.

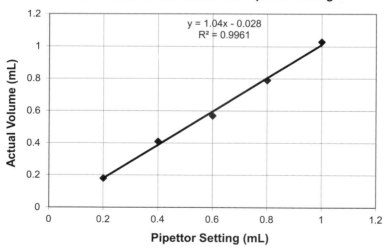

FIGURE 1.6   Sample graph from sample data for Part C.

References

Martin, J.A. 2001. The Art of the Pipet. HMS Beagle, pp.100.
Weast, R.C., (Ed.), 1981–1982. CRC Handbook of Chemistry and Physics, 62nd ed. F-4.

2

Serology

OBJECTIVE

To learn how to perform presumptive tests for blood, semen, saliva, urine, and feces.

SAFETY

Avoid skin contact with solutions. Ensure that this lab is conducted only in an environment with proper equipment and adequate ventilation. Treat all samples with extreme caution; serological evidence, like all biological materials, can be extremely dangerous and hazardous to human health. Always wear filtration masks, gloves, ultraviolet (UV) eye protection, and protective clothing when spraying luminol solutions. Never look directly into the UV light source or allow light beams to bounce off surfaces into your eyes or the eyes of other persons in the vicinity when using alternate light sources. Picric acid is a poisonous, trinitroaromatic compound that is a flammable solid when purchased wet. Picric acid is a high-powered explosive when allowed to dehydrate. As an explosive, picric acid is not shock sensitive, but when in contact with metals, it can form shock sensitive metal picrates. Mercuric chloride is extremely toxic and may be fatal if inhaled, swallowed, or absorbed through the skin. Use in the hood with gloves. Zinc chloride is toxic. Inhalation of and skin contact with zinc chloride should be avoided.

MATERIALS

1. Autoclave
2. Balance
3. Digital camera
4. Hot plate
5. Microcentrifuge
6. Microscope/stereoscope
7. Luminol reagent (0.1 g 5-amino-2,3 dihydro-1,4-phlalazinedione, 5 g anhydrous sodium carbonate in warm water; fill to 100 mL total volume, then add 1.3 g sodium perborate*$4H_2O$, good for use for 6 hours)
8. ABA Card Hematrace blood test (Abacus Diagnostics)
9. ABA Card p30 cards (Abacus Diagnostics)

10. Blood or livers of cows or sheep
11. Disposable plastic pipettes
12. 95% EtOH for clean-up
13. Sterile water
14. Phadebas tablets or paper
15. Hemastix strips
16. Hemident
17. Christmas tree stains (prepared A and B)
18. Kastle-Meyer phenolphthalein reagent in dropper bottles (100 mL of distilled water, 2 g of phenolphthalein, 20 g of KOH, and 10 g of zinc dust; boil until no longer pink, reflux/ stir for 2 to 3 hours, cool to room temperature, and store at 4° C; before use, dilute stock with an equal volume of ethyl alcohol (or one part phenolphthalein and four parts EtOH), store in an amber bottle, good for use for 6 hours after prepared)
19. UV alternate light source (ALS) (and goggles)
20. Cotton swabs
21. Ketchup, lipstick, potato, horseradish, tomato, tomato sauce, red onion, shoe polish, 10% cupric sulfate, 10% nickel chloride, 10% ferric sulfate, as negative controls (Table 2.2)
22. 3% H_2O_2 in a dropper bottle
23. Wood toothpicks
24. 10% $HgCl_2$ (0.1 g of mercuric chloride in 10 mL of ethanol)
25. 10% $ZnCl_2$ (0.1 g of zinc chloride in 10 mL of ethanol)
26. Acid phosphatase test (reagent 1: 10 mL of glacial acetic acid, 12 g of anhydrous sodium acetate, 100 mL of water, 1 g of Brentamine Fast Blue B/Fast Black; reagent 2: 10 mL of deionized water, 0.8 g of alpha naphthyl phosphate; combine 10 mL of reagent 1, 1mL of reagent 2, 89 mL of water, and store in amber bottle for use up to 14 days)
27. Kernechtrot/Picroindigocarmine (KS/PICS) stains and semen sample
28. Blood, semen, saliva, urine, and feces positive control and questioned stains (Table 2.1)

BACKGROUND

Before crime scene investigators (CSIs) can recover evidence from crime scenes. It is important that they properly identify samples that may link the perpetrator to the crime scene or victim. Any evidence with biological material may contain DNA sufficient in quantity for DNA profiling. Blood, semen, saliva, urine, and feces are all valuable potential sources of human DNA evidence that can be presumptively identified as such using serological tests to verify their identity at the crime scene prior to collection (James and Nordby, 2005).

Presumptive tests are qualitative tests that indicate the presence of a substance or class of substances but do not indicate the quantity of the material or its overall quality for use in subsequent analysis. For example, a presumptive test for blood will only verify the presence of blood but not the quantity or quality of the DNA contained in the stain. The presumptive tests in this lab will be used to test for the presence or absence of biological material to help the CSI to decide what to collect and submit to the lab. In this lab, the chemical attributes of the serological materials (Table 2.1) will be used to identify the presence and type of serological material in the simulated evidence samples (Saferstein, 2007).

TABLE 2.1 Chemical Compositions of Body Fluids and Materials and Presumptive Tests

Biological Material	Chemical Composition	Chemical Presumptive Test(s)
Blood	Contains proteins hemoglobin, fibrinogen, albumin, and immunoglobulins and electrolytes, iodine, sulfate, and glucose and other sugars in the liquid plasma; red blood cells (erythrocytes) white blood cells (leukocytes) and platelets (thrombocytes) in the serum	Kastle-Meyer, Hemastix, Hemident/ McPhail's, Hematrace, Hexagon OBTI, RSID, Luminol, Fluorescein, TMB
Saliva	Consists of 99% water containing a complex mixture of aqueous proteins, cells, salts, and sugars; proteins include alpha-amylase, aquaporin, peroxidases, immunoglobulins, lysozyme, the glycoprotein osteonectin, an acidic proline-rich protein, and partially degraded proteins; bacteria, buccal epithelial cells, cystatins, statherin histatins, glucose, o-glycosylated mucins, and cortisol and salt electrolytes, including thiocyanate, sodium, potassium, calcium, magnesium, bicarbonate, and phosphates are also found in saliva; also contains urea, ammonia and histidine-rich peptides	Phadebas, Starch-iodine, SALIgAE-saliva, RSID-saliva test, Amylose Azure, Rapignost-Amylase, Amylase Radial Diffusion Test
Fecal matter	Consists of colon cells, lipids, and indigestible polysaccharides as well as other metabolic wastes	Urobilinogen
Hair	Composed of the proteins collagen (a coiled-coil) and keratin; cuticle, cortex and medulla may vary between individuals and animal species	Microscopy
Urine	Contains nitrogenous waste including urea and uric acid; also contains B-vitamins, electrolytes including sodium, chloride and potassium ions, sulfate, and calcium phosphate, and trace amounts of protein (including amylase, hemoglobin, myoglobin, and Tamm-Horsfall glycoprotein) and organic compounds including bilirubin, creatinine, citric acid, cortisol, dopamine, epinephrine, glucose, homovanillic acid, ketones, porphrins, red blood cells, amino acids, and vanillylmandelic acid	Creatine, BFID-urine, Urease/bromothymol blue

(*Continued*)

TABLE 2.1 Chemical Compositions of Body Fluids and Materials and Presumptive Tests—cont'd

Biological Material	Chemical Composition	Chemical Presumptive Test(s)
Semen	Rich in acid phosphatase, prostate specific antigen, citric acid, inositol, calcium, zinc, magnesium, fructose, ascorbic acid, and prostaglandins; also contains L-carnitine, fructose, neutral alpha-glucosidase, choline, spermine, seminogelin, urea, ascorbic acid, immunoglobulins, and sperm cells; glycolytic and other metabolites and lactic acid; albumin makes up about one third of the protein content of semen and the amino acid content is higher than that of plasma	Phosphatesmo KM Paper Test, PSA (P30), SMITEST, p30, luminol, FastBlue, Christmas Tree Stain, Sodium thymolphthalein
Tears	Consists of proteins, lipids and mucins; wax and cholesterol esters compose 45% of the tear lipid, similar to the meibum while non-esterified cholesterol and phospholipid compose 15% each of the tear lipid; proteins include sIgA, lactoferrin, TSPA, serum albumin, protein G, and the antimicrobial agent lysozyme	
Vaginal mucus	Contains prostaglandins, immunoglobulins (e.g. IgG, IgA), glucose, galactose and fucose sugars, chloride, N-acetylglucosamine, N-acetylgalactosamine, sialic acid, acid phosphatase, lactic acid, citric acid, acetic acid, urea, a peptidase, epithelial cells, pyridine, squalene, a L-fucosidase protein and bacteria and fungi	
Earwax (cerumen)	Consists of long chain saturated and unsaturated fatty acids, alcohols, wax esters and cholesterol, squalene, triglycerides, sugars including galactosamine and galactose, amino acids, keratin, desquamated keratinocytes and hair	
Nasal mucus	Contains proteins and glycoprotein and bacterial cells	
Sweat	Contains sodium chloride, potassium and nitrogen metabolites urea, ammonia and uric acid, lactic acid, chloride, and immunoglobulins	

TABLE 2.1 Chemical Compositions of Body Fluids and Materials and Presumptive Tests—cont'd

Biological Material	Chemical Composition	Chemical Presumptive Test(s)
Breastmilk	Contains essential amino acids, fatty acids, oligosaccharides, calcium, magnesium, phosphorous, zinc and long-chain polyunsaturated fatty acids especially docosahaexaenoic acid (DHA) and anti-infectives including penicillins, cephalosporins, antituberculars, quinolones, sulfonamides, tetracyclines, macrolides, aminoglycosides and antimalarials	

Ideally, the best presumptive tests are not destructive to DNA and are rapid, simple, require only a small amount of material, and do not require the use of any hazardous chemicals. Unfortunately, no currently available presumptive test meets all of these criteria. Always refer to the Material Data Safety Sheet (MSDS) for each chemical with which you work to determine its hazards. Presumptive tests rely on chemical and biological methods to determine the presence or absence of an antigen, protein, or other chemical substance in the stain. Table 2.1 lists the chemical compositions of human body fluids and materials, including those that are probed in serological tests.

Blood is one of the most significant and frequently encountered types of physical evidence associated with forensic investigation of death and violent crime. Table 2.2 shows some common look-alike compounds for body fluids that are often employed as presumptive negative controls in presumptive tests.

CSIs can use portable instruments to run many presumptive tests. These instruments generally do not require chemicals and are rapid, nondestructive, and noninvasive. A UV-light or other alternative light source (ALS) is nondestructive, does not require any consumable reagents, and is often used first in a crime scene investigation. The ALS causes blood to fluoresce and also may be used to visualize saliva, semen, urine, and feces on the surface of materials due to their fluorescence. Raman and attenuated total reflectance Fourier transform-infrared (ATR FT-IR) spectroscopy can be used to differentiate materials and to identify saliva, urine, semen, and sweat. Portable, handheld instruments are now available (Virkler and Lednev, 2009).

Additional instrumental methods include light microscopy and scanning electron microscopy (SEM-EDX). SEM-EDX can identify saliva, urine, semen, and sweat, but samples must be sent to the lab to utilize these methods. Microscopy is the preferred presumptive test for hair examination including species differentiation (Virkler and Lednev, 2009).

Serological tests may evaluate the presence of biological material by employing antibody-antigen reactions. These can be considered confirmatory tests for the presence of human body fluids. Species determination of blood may be performed using the ABA HemaTrace (Abacus Diagnostics, West Hills, California), Hexagon OBTI (BLUESTAR Forensics), HemeSelect (Spear and Binkley, 1994), or RSID Blood (Independent Forensics) immunochromatographic tests. These antibody tests, first validated in 1999, are rapid, highly selective methods of testing biological materials and determining species, but their use requires

TABLE 2.2 Look-alike Substances That Are Presumptive Negatives for Biological Material

Biological Material	Look-alike substance (presumptive negative)
Blood	Barbeque sauce, pink lotion, lipstick, wine, ketchup, nail polish, Catalina dressing, chocolate
Saliva	Water, clear drinks
Fecal matter	Chocolate
Hair	Animal hair, animal fur, threads
Urine	Apple juice, Italian dressing, coffee
Semen	Italian dressing, gels
Tears	Water, clear drinks
Vaginal mucus	Yogurt, lotions/gels, cream cheese, mayonnaise, petroleum jelly
Earwax (cerumen)	Candle wax
Nasal mucus	Hair gel
Sweat	Water, clear drinks
Breastmilk	Cow milk

unique reagents and they are relatively expensive (Hochmeister et al., 1999a, 1999b). Although the RSID Blood test has been shown to be specific for human blood, the ABA HemaTrace will react with blood from humans, higher-order primates, and ferrets. Older immunological species origin determination techniques include immunoelectrophoresis, single and double diffusion using agar plates, and hemagglutination, a test to visualize the antibody-antigen reaction clumping. Instrumental methods including UV-vis and fluorescence spectroscopic techniques have also been applied to blood classification (Virkler and Lednev, 2009, Johnston et al., 2008, Tobe et al., 2007).

Specific presumptive tests are also available to determine the presence of other biological fluids including saliva, semen, vaginal fluid, urine, and feces. The Phosphatesmo KM Paper Test (BVDA, Amsterdam, Holland) for acid phosphatase, PSA (P30) immunochromatographic antibody-antigen test (SMITEST: Seratec Diagnostica, Gottingen, Germany); p30 test (Abacus Diagnostics, West Hills, California), choline oxidase detection via luminol, FastBlue, and the Christmas tree stain (Kernechtrot/Picroindigocarmine) and fluorescence microscopy tests are used to determine the presence of semen. Sodium thymolphthalein monophosphate, the zinc-paper strip test, Barberio test, and Puanen's test may also be used to detect a seminal stain. The alternative light source, capillary electrophoresis (CE), high-performance liquid chromatography (HPLC), and thin layer chromatography (TLC) have been used to detect semen (Virkler and Lednev, 2009).

The creatine, Nessler's, DMAC, urease/bromothymol blue, Jaffe, Salkowski, and TZ-UA one/Urea NB tests are used to detect the presence of urine chemically. Immunological tests for urine include the Sandwich ELISA and SP radioimmunoassay. Chromatographic tests including HPLC, electrospray ionization-liquid chromatography-mass spectrometry

(ESI-LC-MS), PC, and TLC have been used to evaluate urine samples. The BFID-urine kit and ALS are used to detect the presence of urine in some labs (Virkler and Lednev, 2009).

Saliva is detected by Phadebas tablets for amylase (Figure 2.1), the SALIgAE-saliva test (Abacus Diagnostics, West Hills, California), the RSID-saliva test (Independent Forensics, Hillside, Illinois), the starch-iodine mini-centrifuge test, starch-iodine test, Amylose Azure, Rapignost-Amylase, or the Amylase Radial Diffusion test. The SALIgAE-saliva test (Abacus Diagnostics, West Hills, California) is a nonamylase colorimetric test (Virkler and Lednev, 2009).

The presence of fecal matter is evaluated by the urobilinogen (Edelman's) test. Vaginal fluid is detected by the ALS, PAS reagent, starch gel electrophoresis, capillary isotachophoresis, and immunologically using the estrogen receptors. Antihuman alpha-lactalbumin antibody is used to probe the presence of breast milk clinically. There is no published presumptive forensic test for breast milk, tears, or nasal mucus (Virkler and Lednev, 2009).

The invasive chemicals and swabbing often required for presumptive tests can present CSIs with considerable challenges. Chemicals may damage or destroy chemical associative trace evidence and render a sample unusable for genetic testing. Evidence suspected of containing biological material may be swabbed or a moistened filter paper may be pressed on the questioned area(s). These methods allow the CSI to evaluate large areas without applying the chemical directly to the evidence (James, Kish and Sutton, 2005). The location of the biological material can be determined, collected, and preserved for further testing. The postprocessing often necessitates the amplification of the biological material via the polymerase chain reaction (PCR) to detect the presence at a significant cost. CSIs select methods that will not compromise biological evidence for further DNA testing first and then employ more invasive methods only if necessary. For example, Hemastix strips have been reported to interfere with subsequent sample processing with the DNA IQ system. Hemident and leucomalachite green inhibited DNA recovery in a recent study.

The blood tests to be performed in this experiment include the Hemident, Kastle-Meyer, Hemastix, luminol, alternate light source, and ABA HemaTrace tests.

Leucomalachite green is the reduced or colorless form (leuco) of the malachite green dye. It is also referred to as McPhail's Reagent (Hemident) (Figure 2.2). Leucomalachite green

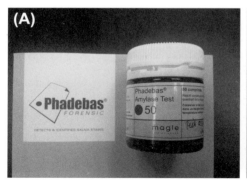

FIGURE 2.1   A. Phadebas test tablets and filter paper for saliva. B. Positive Phadebas test for saliva.

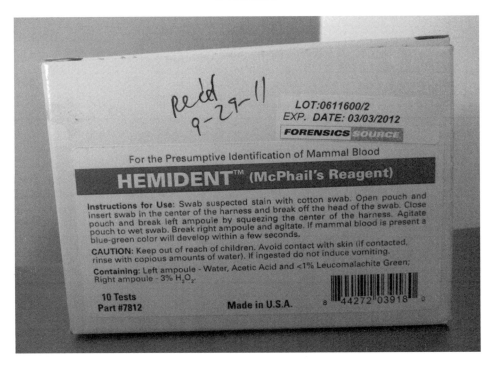

FIGURE 2.2   Hemident kit.

oxidation is catalyzed by heme in blood to produce a blue-green color (positive result). The reaction is often carried out in acid (acetic acid) with hydrogen peroxide as the oxidizer, although some procedures use sodium perborate. A substance that is not blood will produce no color change. False negatives with similar redox properties have been identified. Use of this chemical has been shown to obstruct DNA recovery from the stain.

The Kastle-Meyer color test for blood employs the acid-base reaction of phenolphthalein. Chemically, blood hemoglobin possesses peroxidase-like activity because of the heme. Peroxidases are enzymes that accelerate the oxidation of several classes of organic compounds by peroxides. When a bloodstain, phenolphthalein reagent, and hydrogen peroxide are mixed together, the blood's hemoglobin will cause the formation of a deep pink color (positive result) (Figure 2.3). A material lacking peroxidase activity will not cause the formation of the characteristic deep pink color in the presence of the reagents; however, potatoes and horseradish have been shown to give false positive results.

Field investigators have found Hemastix strips to be another useful presumptive field test for blood and urine (Figure 2.4). Like the Kastle-Meyer test, Hemastix strips will react with peroxidases. The Hemastix are 3-inch plastic strips with a filter paper containing a proprietary mixture of blood reagent chemicals at the tip. These include TMB, diisopropylbenzene dihydroperoxide, buffering materials, and inactive materials. Designed as a urine dipstick test from blood, the yellow strip can be moistened with distilled water and placed in contact with the suspect bloodstain. (It is important not to put the Hemastix

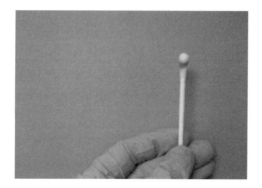

FIGURE 2.3   Positive Kastle-Meyer test.

strips on the stain, as they have been shown to inhibit downstream DNA typing.) The appearance of a green or blue-green color (read by comparison to the chart on the bottle) indicates blood (Figure 2.4, A-C). Swabs pretreated with cosmetic substances may yield false positive results.

The luminol reagent (0.1 g 5-amino-2,3 dihydro-1,4-phlalazinedione, 5 g sodium carbonate in 100 mL of distilled water and 0.7 g of sodium perborate added before use) luminesces in the presence of blood. The iron from the hemoglobin heme group in the blood serves as a catalyst for the chemiluminescence reaction that causes luminol to oxidize, so a blue-white or yellow-green glow is produced when the solution is sprayed where there is blood. Only a tiny amount of iron is required to catalyze the reaction: blood stains diluted 1:5,000,000 can be detected. The blue glow lasts for about 30 seconds before it fades, which is enough time to take photographs of the areas so they can be investigated more thoroughly. Extra sprayings will often result in stain pattern diffusion. The luminol test has been shown to not adversely affect the polymerase chain reaction used subsequently to amplify DNA for DNA typing, but it has been shown to falsely react with peroxidases and metals (Webb et al., 2006, Sidorov et al., 1998).

Fluorescein ($\lambda_{ex\ max}$ = 494 nm, $\lambda_{em\ max}$ = 521 nm, in water) is also used to detect the presence of blood and is prepared much like the phenolphthalein reagent. Fluorescein is reduced in alkaline solution over zinc to fluorescin, which is then applied to the suspected blood-stained area. The catalytic activity of the heme then accelerates the oxidation by hydrogen

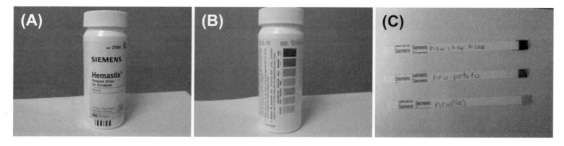

FIGURE 2.4   A. Hemastix reagent test strips. B. Hemastix color matching chart. C. Hemastix positive (top), false positive (middle) and negative reactions (bottom).

peroxide of the fluorescin to fluorescein that will then fluoresce when treated with UV light. In commercial preparations, a thickener is added that makes the solution adhere to the surface and is thus much more effective on vertical surfaces. Once the mixture is sprayed, visualization of the fluorescence requires the use of an ALS, typically set to 450 nm. Fluorescein, unlike luminol, will not work in the presence of household bleach. No degradation of the DNA has been observed after the application of this reagent.

Microcrystalline tests such as the Takayama and Teichmann can also be used to indicate the presence of blood and are considered to be confirmatory tests for that substance. In the Takayama test, if heme is gently heated with pyridine under alkaline conditions in the presence of a reducing sugar such as glucose, crystals of pyridine ferriprotoporphyrin or hemochromogen are formed. The best results were found with a reagent containing water, saturated glucose solution, sodium hydroxide (10%), and pyridine in a ratio of 2:1:1:1 by volume. A small stain is placed under a coverslip and the reagent is allowed to flow under the coverslip to saturate the sample. After a brief heating period, the crystals are viewed microscopically. Using a very small (2 mm square or smaller) coverslip allows the test to be carried out on a small stain quantity.

Species-specific immunochromatographic tests evaluate characteristics unique to blood, including antigens present on the surface of red blood cells. Abacus Diagnostics ABA Hema-Trace Cards (Figure 2.5) are rapid, sensitive, and accurate immunochromatographic antibody-antigen tests that use a high-titer antihuman hemoglobin serum and thus are somewhat species specific. The positive reaction will be observed with blood from higher-order primates, including humans, and ferrets (Figure 2.6). The detection limit is 0.07 µg Hb/mL.

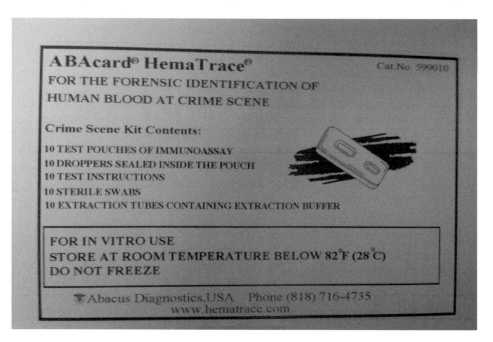

FIGURE 2.5   Abacus Diagnostics, Inc. ABA Card Hematrace kit.

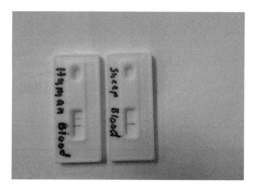

FIGURE 2.6 ABA Card Hematrace positive (left) and negative (right) reactions.

The antihuman hemoglobin and human hemoglobin combine to form a complex that is immobilized to the T line and is visualized by a conjugated pink dye. As an internal control an anti-immunoglobin antibody is immobilized at the C line and is visualized by the same dye when the antihuman hemoglobin binds. The test is invalid if the C line does not appear. In a positive result, both the C and T lines are present, whereas the negative result will show only the C line present (Figure 2.6).

Acid phosphatase is an enzyme found throughout the body, but it is produced in high quantities by the prostate gland and thus seminal fluid contains high concentrations of this enzyme (Yakota et al., 2001). This test can indicate the presence of a male even in cases where the male is aspermic. The test utilizes the activity of this enzyme. Abacus Diagnostics produces a P30 test (Figure 2.7) for semen that works in the same manner as the species-specific blood tests. It is an immunochromatographic antibody-antigen tests that use a high-titer antihuman prostate 30 antigen serum.

The Christmas tree (aluminum sulfate, nuclear fast red, picric acid, indigo carmine) (Kernechtrot/Picroindigocarmine) stain test is a test for the presence of spermatozoa cells as viewed using a compound light microscope. This reagent stains the sperm heads pink to red and the tails green, hence the name. Cells are fixed to the microscope slide with heat. False negative results are produced from azoospermic/oligospermic/vasectomized males. Sodium thymolphthalein monophosphate may also be used to detect a seminal stain. A positive result is indicated by the presence of a deep blue color, whereas the formation of a green to yellow color is a negative result. A new stain, the sperm HY-Liter, was developed using sperm-specific antibodies to make the sperm fluoresce so they can better be differentiated from female epithelial cells.

The creatine test used to detect the presence of urine requires picric acid and 5% NaOH. A positive result is indicated by the presence of an orange color, whereas the formation of a yellow color is a negative result. The presence of fecal matter is tested by the urobilinogen test. The test uses 10% mercury chloride and 10% zinc chloride solutions with the result viewed under UV light. A positive result is indicated by the presence of a stable, apple-green fluorescent color.

In this laboratory experiment, biological materials will first be detected using an ALS as it requires no chemicals and is non-destructive. Subsequently, blood will be identified using the

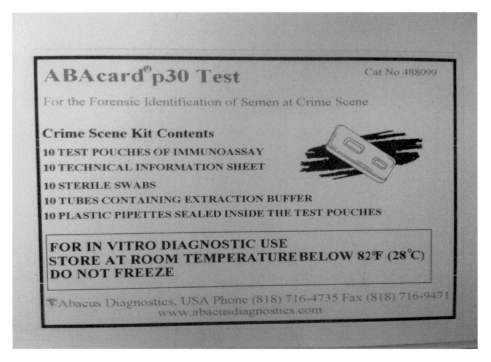

FIGURE 2.7   Abacus Diagnostics p30 test.

luminol, Hemastix, Hemident, Kastle-Meyer and ABA card tests. Semen will be detected using the acid phosphatase, and ABA Card p30 tests and sperm will be visualized using the Christmas Tree Stains and Teichmann microcrystalline test. Saliva will be detected using the Phadebas tablets or paper and urine will be detected using the creatine test. Finally, the urobilinogen test will be used to probe for the presence of fecal matter.

PROCEDURE

Numbers (measurements) should *always* be accompanied by units!

Part A: Specimen Handling

1. Examine one item at a time over clean butcher paper or similar material.
2. Unwrap items one at a time over the working surface; collect any material of evidential value (e.g., hairs, fibers, trace material).
3. Visually examine the item.
4. Test all presumptive reaction solutions on the day of use with positive control samples to ensure that the solutions are working properly. Document all positive and negative controls.
5. Collect and package all stains and samples with positive results for further testing.

Part B: Detecting the Presence of Biological Fluids Using an Alternate Light Source

1. Turn on the alternative light source (UV or visible) and allow it to warm up.
2. Darken the room where the examination is to occur.
3. Use UV goggles for viewing.
4. Systematically scan the light at a 45-degree angle over the evidence and look for stains, which fluoresce, or other evidence that may be of value. Fluorescence indicates a positive result.
5. Note the stain areas for future testing by circling them with a pencil.
6. Semen stains will not always fluoresce due to substrate or concentration. If no stains appear, it may be necessary to examine for stains either visually or with swabbing. Urine, saliva, vaginal secretions, sweat, liquids, food, and the like all may fluoresce. Test with positive controls to observe the expected response. Proceed with other presumptive tests to determine the presence of each type of biological material.

Part C: Chemical Blood Tests

a. Luminol

1. Obtain both positive and negative control samples and unknown samples.
2. Prepare the luminol reagent if this has not already been prepared. The solution is very unstable; do not mix it in advance. Decant the solution into a plastic spray bottle with no metal parts.
3. In a dark area, spray a fine mist over areas to be tested. A positive result will yield an intense and prolonged production of blueish light (chemiluminescence lasting approximately 30 seconds). A negative result will be indicated by the absence of luminescence. An inconclusive result is indicated if the reaction does not conclusively reflect either the positive or negative result. Note that luminol will also react with metals, cleaning solutions, and other surfaces yielding false positive results.

b. Hemastix Test (Multitest for Urine/Blood)

1. Obtain a Hemastix strip. Add one drop of distilled water to the strip. The test pad should remain yellow in color.
2. Obtain both positive and negative control samples and unknown samples. Add one drop of blood to the strip or swap moistened with the stain. A green color indicates a positive result, as noted on the side of the reagent bottle. A negative result is indicated by the yellow color of the test strip or no color change. A false positive can be observed with potato or horseradish.

c. Phenolphthalein (Kastle-Meyer)

1. Swab or take a small cutting of the stain or area in question by swabbing the stain with a moist (use distilled water) cotton swab.
2. Add one drop of the phenolphthalein working solution to sample.

3. Add one drop of 3% hydrogen peroxide solution to sample.
4. Read the results immediately. A positive test result will yield a pink color. The negative result will be indicated by no color change. The result is marked inconclusive if the reaction does not conclusively reflect either the positive or negative result. Metal ions capable of redox reactions will yield false positive results.

d. Immunochromatographic Test (OneStep ABAcard HemaTrace, Abacus Diagnostics, Inc.)

1. Extract a portion of the stained material in an appropriate amount of the extraction buffer provided in the test for up to 30 minutes. (Older stains can be extracted in a 5% ammonium hydroxide solution. Following the extraction and centrifugation, allow the ammonium hydroxide to evaporate and then add 150 µl of the extraction buffer provided in the test.)
2. Apply the diluted extract to the sample well of the test strip.
3. Read at 10 minutes, but do not read after 10 minutes.
4. A positive result will be indicated by the formation of a line in the test (T) and control (C) region of the membrane within 10 minutes. A negative result will be indicated by the formation of a line only in the C region at 10 minutes. An inconclusive result will be indicated by a reaction that does not conclusively reflect either the positive or negative result. The test is ruled invalid if the line in the C region is still absent at 10 minutes.

Part D: Semen Chemical Tests

a. Acid Phosphatase (AP)

1. Moisten filter paper or stain with sterile deionized H_2O. Press moistened filter paper to stain (overlay). If stain has been moistened, use dry filter paper. If a swab or filter paper will not work, cut a small portion of the stain.
2. Remove filter paper and apply working AP solution to the filter paper, swab, or cutting.
3. Read results at 1 minute.
4. A positive result is indicated by the presence of a purple color, and a negative result is indicated by the absence of purple color within 1 minute. The test is considered inconclusive if the reaction does not decisively reflect either the positive or the negative result.

b. Kernechtrot/Picroindigocarmine or Christmas Tree Stain for Spermatozoa

1. Add a swab sample to slide, dry, and heat fix.
2. Add KS (A) to the fixed slide for approximately 15 minutes.
3. Rinse with deionized H_2O.
4. Add PICS (B) to slide for approximately 15 seconds.
5. Rinse with ethanol.
6. View the stained slide using the compound light microscope. A positive result is indicated by the presence of a sperm head stained two-toned red with green sperm tails, and a negative result is indicated by the absence the red and green colors (no spermatozoa

observed). An inconclusive result is a stain that contains material that cannot be conclusively identified as spermatozoa.

Rating system (magnification at 400×):

+1 = 1 to 10 sperm/slide
+2 = 1 sperm/every 6 fields
+3 = 1 sperm/every 3 fields
+4 = 3 or more sperm/field

c. P30 Immunochromatographic Test (Seratec PSA Semiquant or Abacus Diagnostics P30 Test)

1. Extract a portion of the seminal stain in approximately 500 μl of deionized water or extraction buffer in kit.
2. Microcentrifuge the extract from the material.
3. The sample should be at room temperature.
4. Add extract (approximately 200 μl or five drops) to the test well.
5. Read and record results.
6. Do not read after 10 minutes. A positive result is indicated by the formation of a line in both the result (T) region and the control (C) region of the membrane within 10 minutes. If a line in the C region does not appear at 10 minutes, the result is negative. If the reaction does not conclusively reflect either the positive or negative result, the test is inconclusive. If a line in the C region at 10 minutes does not appear, the result is inconclusive.

d. Teichmann Microcrystalline Test (Modified by Symons, 1913)

1. Dissolve sodium iodide to 1% in strong lactic acid. This solution becomes brown over time, but the change does not alter its action.
2. Cover the dried stain (use a heating plate) with the solution, place the cover glass in position on top, and warm slowly until the solution is just about to boil under the cover glass for 5 minutes or until characteristic crystals appear. It may be necessary to add more solution under the cover glass after 5 minutes of heating.
3. View using a microscope. Characteristic Teichmann crystals indicate a positive result, and no Teichmann crystals indicate a negative result.

Part E: Saliva Test

Amylase Detection Using Phadebas Tablets

1. Extract a cutting of the sample in 100 μl of distilled or deionized water.
2. Pipette approximately 5 μl of the extract into a tube.
3. Add approximately 150 μl of the Phadebas stock solution or crushed tablet or apply to Phadebas filter paper.
4. Incubate at approximately 37° C for 30 minutes.

5. Centrifuge and read immediately. A dark blue supernatant indicates a positive result, whereas a clear supernatant indicates a negative result. The test is inconclusive if a light blue supernatant is observed. With the Phadebas paper, the breakdown and loss of blue color is indicative of a positive result and intact blue color is indicative of a negative result.

Part F: Urine Test

Creatine Test for Urine

1. Place the stained material (approximately 2-mm square) on a small filter paper.
2. Add one drop of saturated aqueous picric acid (IN HOOD).
3. Add one drop of 5% NaOH.
4. Allow to stand at least 15 minutes.
5. Examine coloration. An orange color indicates a positive result, whereas a yellow color indicates a negative result. The test is inconclusive if neither the positive nor the negative result is reflected.

Part G: Feces Test

Urobilinogen Test for Fecal Matter

1. Extract a portion of swab or stain in 200 μl of sterile deionized water for approximately 15 minutes.
2. Pipet two drops extract into a tube.
3. Add three drops 10% $HgCl_2$ and mix.
4. Add three drops 10% $ZnCl_2$ and mix.
5. Examine the liquid under UV light or an alternate light source (leaving the sample at room temperature for approximately 30 minutes may enhance the color). A positive result is indicated by a stable apple-green fluorescence indicates a positive result, whereas the absence of the fluorescence indicates a negative result. The test is inconclusive if neither the positive nor the negative result is produced.

QUESTIONS

Tabulate all results, including those with positive and negative controls for each test. Report results as positive (inclusion), negative (exclusion), or inconclusive. Did any samples not produce the expected results? State the reasons.

References

Hochmeister, M.N., Budowle, B., Rudin, O., Gehrig, C., Borer, U., Thali, M., Dirnhofer, R., 1999. Evaluation of prostate-specific antigen (PSA) membrane test assays for the forensic identification of seminal fluid. J. Forensic Sci. 44, 1059−1060.

Hochmeister, M.N., Budowle, B., Sparkes, R., Rudin, O., Gehrig, C., Thali, M., et al., 1999. Validation studies of an immunochromatographic 1-step test for the forensic identification of human blood. J. Forensic Sci. 44, 597−602.

James, S.H., Kish, P.E., Sutton, T.P., 2005. Principles of Bloodstain Pattern Analysis: Theory and Practice, third ed. CRC Press, Boca Raton, FL, p. 542.

James, S.H., Nordby, J.J., 2005. Forensic Science: An Introduction to Scientific and Investigative Techniques, second ed. CRC Press, Boca Raton.

Johnston, E., Ames, C.E., Dagnall, K.E., Foster, J., Daniel, B.E., 2008. Comparison of presumptive blood test kits including hexagon OBTI. J. Forensic Sci. 53, 687–689.

Saferstein, R., 2007. Basic Laboratory Exercises for Forensic Science; Exercises 9-10 & 31. Pearson Prentice Hall, Upper Saddle River, NJ, p. 389.

Sidorov, V.L., Maiatskaia, M.V., Smolianitskiĭ, A.G., Babakhanian, R.V., 1998. [Establishing the presence of blood in stains for material evidence by using the luminescent blood test][Russian]. Sud. Med. Ekspert. 41, 20–23.

Spear, T.F., Binkley, S.A., 1994. The HemeSelect test: a simple and sensitive forensic species test. J. Forensic Sci. 34, 41–46.

Tobe, S.S., Watson, N., Nic Daeid, N., 2007. Evaluation of six presumptive tests for blood, their specificity, sensitivity, and effect on high molecular-weight DNA. J. Forensic Sci. 52, 102–109.

Virkler, K., Lednev, I.K., 2009. Analysis of body fluids for forensic purposes: from laboratory testing to non-destructive rapid confirmatory identification at a crime scene. Forensic Sci. Int. 188, 1–17.

Webb, J.L., Creamer, J.I., Quickenden, T.I., 2006. A comparison of the presumptive luminol test for blood with four non-chemiluminescent forensic techniques. Luminescence 21, 214–220.

Yakota, M., Mitani, T., Tsujita, H., Kobayashi, T., Higuchi, T., Akane, A., Nasu, M., 2001. Evaluation of a prostate-specific antigen (PSA) membrane test for forensic identification of semen. Leg. Med. (Tokyo) 3, 171–176.

3

Sampling Biological Evidence for DNA Extraction

OBJECTIVE

To learn how to identify, collect, handle, package, and label biological evidence items.

SAFETY

Place biohazards in the biohazard bag as indicated by your instructor for autoclaving.

Biological materials have the potential to transmit infectious diseases; handle all samples with extreme caution. For more information, refer to Biosafety in Microbiological and Biomedical Laboratories, U.S. Department of Health and Human Services, 1993, and OSHA Bloodborne Pathogen Standard 29 CFR, part 1910.1030, U.S. Department of Health and Human Services.

MATERIALS

1. Simulated evidence samples (e.g., letters with stamp, cigarette butts, soda can, chewing gum, human hair with root, fingerprint, saliva/buccal sample, adhesive bandage with blood, hat with headband, dress shirt with collar)
2. Gloves, lab coats, and goggles
3. 1X Tris-EDTA (TE) extraction buffer (10 mM Tris-0.1 mM EDTA, pH 8) or 0.1X TE buffer or extraction buffer provided in selected DNA extraction kit—for example, QiaAMP (Qiagen), DNAIQ (Promega), PrepFiler (Applied Biosystems)
4. Sterile microcentrifuge tubes

BACKGROUND

Crime scene investigators (CSIs) are responsible for recovering evidence from crime scenes. It is important that they properly identify, collect, handle, package, and label evidence

27

items recovered from the crime scene. Most important, CSIs must carefully document the location of evidence at the crime scene through sketches and photographs.

Any biological evidence may contain DNA. Blood, semen, saliva, urine, feces, vaginal fluid, earwax, fingerprints, fingernails, muscle tissue, bone, hair, dandruff, mucus, sweat, human milk, and even fingerprints are all valuable potential sources of DNA evidence (Bond and Hammond, 2008, Drobnic, 2003, Lorente et al., 1998). This material may be found on or impregnated in almost any surface, including condoms, fingerprints, blood spatter, cigarette butts, drinking glasses, tape, aluminum cans, stamps, toothpicks, skin tissue, cotton swabs, clothing, eyeglasses, masks, objects, weapons, chewing gum, bullets, fingernail scrapings, and bite marks (Bond and Hammond, 2008, Drobnic, 2003, Lorente et al., 1998). Blood spatter and other questionable stains can be checked using serological tests to verify their identity at the crime scene prior to collection. Examples of surfaces that may be holding DNA-containing evidence and may be collected from a crime scene are shown in Table 3.1 and Figure 3.1.

In 1991, Dr. Henry Lee wrote about how DNA can be directly transferred and listed eight types of DNA that law enforcement personnel should be concerned with:

1. The suspect's DNA deposited on the victim's body or clothing
2. The suspect's DNA deposited on an object

TABLE 3.1 Types of Evidence Found at Crime Scenes

Biological fluid or material	Examples of evidence sources
Saliva	Bite mark, soda can, beer bottle, cigarette butt, used stamp, envelope seal, toothpick, bedding
Blood	Clothing, carpet, floor, victim, bullet, walls, bedding
Hair	Fabric, floor, clothing, bedding
Vaginal fluid	Undergarments, condom, clothing, bedding
Semen	Victim, clothing, undergarments, condom, bedding
Sweat	Hat, bandana, scarf, mask, eyeglasses, clothing
Tears	Clothing, tissue
Mucus	Tissue, cotton swab, clothing
Earwax	Cotton swab, tissue
Urine	Clothing, floor, bed, undergarments
Feces	Undergarments, clothing
Fingerprints	Baseball bat, steering wheel, glass, can, bottle
Skin cells	Fingernail scraping, dandruff
Fingernail	Floor, bedding, clothing
Bone	Body or skeletal remains
Teeth	Body or skeletal remains

FIGURE 3.1 Evidence sources of DNA. *(Data taken from Bond and Hammond, 2008.)*

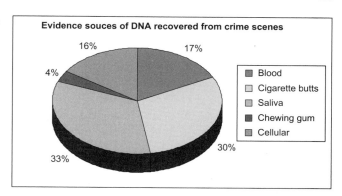

3. The suspect's DNA deposited at a location
4. The victim's DNA deposited on the suspect's body or clothing
5. The victim's DNA deposited on an object
6. The victim's DNA deposited at a location
7. The witness's DNA deposited on the victim or suspect
8. The witness's DNA deposited on an object or at a location

The presence of a positive DNA profile match for a suspect from DNA evidence does not necessarily implicate the suspect in a crime, but it does place the suspect at the crime scene at some point in time.

Modern DNA typing techniques require extremely small amounts of material. As a result, the risk of contamination of the biological evidence is extremely high. For example, one cell is sufficient for DNA typing; however, one sneeze, cough, or even a light touch can deposit more than enough DNA to contaminate the subject evidence. Therefore, CSIs and laboratory personnel should take the following precautions to minimize the risk of contamination (Goray et al., 2012, Rutty et al., 2003):

1. Always wear gloves whenever handling evidence.
2. Ensure gloves are cleaned with a 95% alcohol or 10% bleach disinfectant.
3. Change gloves when handling two separate or different evidence items.
4. Change gloves immediately if they come into contact with the investigator's face, mouth, or nose or knowingly become contaminated.
5. Wear a clean, dedicated lab coat.
6. Wear a face mask to prevent contamination from coughing or sneezing.
7. When possible, use disposable instruments; if disposable instruments are not available, use stainless steel instruments that can be autoclaved, and change or clean instruments between each use.
8. Clean all nondisposable instruments with a 95% ethanol solution or autoclave.
9. Wash lab benches using a 10% bleach solution, followed by a 95% ethanol or 95% reagent alcohol rinse. Place clean paper or absorbent pads on bench tops when the area is in use, and change as necessary. If a lab bench is contaminated, the absorbent pads should be folded inward and placed into an autoclave bag and the bench rewashed with bleach and

ethanol as noted earlier. The autoclave bag should be double-bagged and properly disposed of at the location of the autoclave.

10. Clean lab equipment (test tube rack, instruments, pipettes, thermocyclers, etc.) and supplies by either exposing them to ultraviolet (UV) light for a minimum of 20 minutes per surface area or by using a 10% bleach solution.

11. Allow all biological evidence to completely air dry prior to analysis; if storing moist biological evidence, ensure it is packaged in an air-permeable container/package and sealed using evidence tape or biohazard labeling (as needed).

12. Only store DNA evidence in a temperature-controlled environment; refrigeration of biological sample will help prevent the growth of mold and bacteria. Liquid blood samples should be stored in the freezer ($-20°$ C).

13. Clearly mark all evidence items with the case number, item number, collection date, and initials of the evidence collector on the package seal.

14. Store the victim's and perpetrator's clothing separately from each other.

15. Open only one DNA-containing sample tube at a time.

Finally, CSIs must scrupulously control evidence chain of custody, or the chronological documentation of the possession, custody, control, transfer, analysis, and disposition of the evidence. Good chain-of-custody procedures help avoid contamination as well as ensure the impartiality of the evidence. Access to the evidence must be restricted to the fewest number of authorized personnel.

To determine the probability that a given sample derives from a given individual, CSIs must collect elimination and reference samples. Elimination samples can be collected from any law enforcement and lab personnel with access to the crime scene or evidence and are used to determine whether a given piece of biological evidence originated from the suspect or another known-identity individual. A reference sample is biological evidence collected from a verified or documented source (e.g., suspect, victim, missing person, or family member) that when compared to the evidence collected at the crime scene can show the link between a perpetrator and the crime scene or missing person to himself or herself or his or her family.

If collecting liquid blood for reference samples, collect seven cubic centimeters (cc) of blood (one cc in infants and five cc in children) in the purple top tubes with ethylene diamine tetraacetic acid (EDTA) preservative and refrigerate the samples until they are delivered to the evidence storage facility. Alternatively, collect buccal swabs (Figure 3.2) or use a sterile disposable toothbrush A substrate control collected by using the same sterile distilled water and lot of swabs or surface material (paint, carpet, etc.), as the evidence sample should be collected to compare evidence swabs if water contamination is later suspected.

The CSI assesses, collects, and ships the biological evidence to the forensic laboratory for more sophisticated analysis by a forensic lab analyst (President's DNA Initiative). Upon receipt at the lab, the analyst should do the following:

1. Unpack, inspect, and inventory the evidence on a clean, immobile surface in a secure, dedicated room.

2. Swab the evidence material(s) using a sterile cotton swab moistened with autoclaved, distilled water or cut an appropriate quantity of the sample for use.

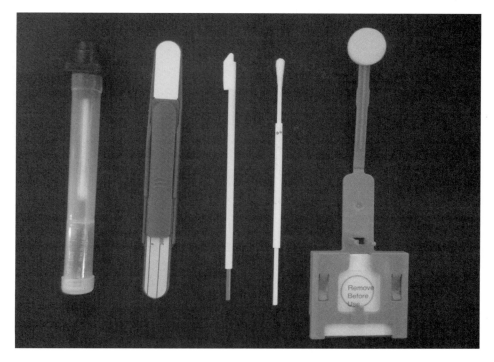

FIGURE 3.2   Sample body fluid collection devices.

3. Store the cotton swab or cutting sample in a sterile, clearly labeled tube (if it is not going to be analyzed immediately).

Following analysis, the lab analyst should replace the evidence in a tube with a new seal. The lab analyst should have only one piece of evidence and one tube open at any one time. After DNA is extracted from the samples, DNA samples should be stored at 4°C or −20°C in the lab (or −70°C for long-term storage).

In this experiment, you will collect (Figure 3.2), package, and label each simulated source of biological evidence as provided by your instructor. Samples should be portioned out so that the sample is not completely used up; triplicate samples are preferred when possible. A 1X Tris-EDTA (10 mM Tris-HCl-0.1 mM EDTA) (TE) buffer will be added to each sample to facilitate the first step of the DNA extraction procedure.

PROCEDURE: DNA COLLECTION AND PACKAGING

Part A: Cigarette Butts

1. Collect a portion of the filter or paper of the cigarette butt in the area that would be in contact with the mouth or hand.
2. For three replicates (preferred), cut into three smaller pieces lengthwise and place each in sterile 1.5-mL microcentrifuge tubes containing spin baskets and 1 mL of 1X TE extraction

buffer (or other desired extraction buffer) from the first step of the desired extraction procedure.

Part B: Chewing Gum

1. Divide the chewing gum (Figure 3.3) into three approximately equal pieces by cutting it with a clean scalpel.
2. Place the cut pieces in separate sterile 1.5-mL microcentrifuge tubes containing spin baskets and 1 mL of 1X TE extraction buffer (or other desired extraction buffer) from the first step of the desired extraction procedure.

Part C: Stamps, Envelopes

1. Carefully open the flap or remove the stamps (Figure 3.4) using steam and clean tweezers.
2. Cut a portion of the sample and place it in a sterile 1.5-mL microcentrifuge tube containing a spin basket. Alternatively, swab the gummed flap or stamp and dissect the swab and add it to a sterile 1.5-mL microcentrifuge tube or cut a portion of the flap or stamp without opening or removing it from the envelope and place it in sterile 1.5-mL microcentrifuge tube containing a spin basket.
3. Add 1 mL of 1X TE extraction buffer (or other desired extraction buffer) from the first step of the desired extraction procedure. Mix thoroughly.

Part D: Tissues (Body, Plant)

1. Remove an approximately 1-cm square piece of tissue.
2. Mince or cut the 1-cm square piece into smaller pieces, and add it to a sterile 1.5-mL microcentrifuge tube.
3. Add 1 mL of 1X TE extraction buffer (or other desired extraction buffer) from the first step of the desired extraction procedure. Mix thoroughly.

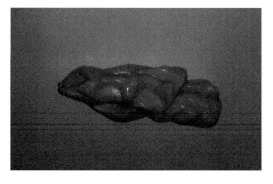

FIGURE 3.3   Chewing gum simulated evidence sample.

FIGURE 3.4   Stamps simulated evidence sample.

Part E: Hair (Unmounted)

1. Use forceps to transfer hair (Figure 3.5) to a piece of paper that can be folded or to an envelope or vacuum with a sterile filter if necessary to collect.
2. Examine hair under a dissecting microscope for the presence of material, noting the possible presence of body fluids that may interfere with typing of hair.
3. Wash the hair by immersing it in sterile saline in a sterile 1.5-mL microcentrifuge tube.
4. Rinse the hair thoroughly in ethanol and place it on clean paper.
5. Cut a 1-cm section from the root end and place in a sterile tube. A 1-cm section of the shaft adjacent to the root may be placed in a separate tube to be utilized as a substrate sample (substrate control) to check for cellular material on the surface of the hair.
6. Add 1 mL of 1X TE extraction buffer (or other desired extraction buffer) from the first step of the desired extraction procedure. Mix thoroughly.

Part F: Liquid Sample (e.g., Blood)

1. Collect wet blood on a marked area of filter paper or FTA card (do not centrifuge), and allow it to dry or for venous blood collection; collect blood in a purple-topped Vacutainer tube containing EDTA anticoagulant and centrifuge cells to pellet.
2. Remove supernatant from the pelleted blood sample, being careful not to disturb the pellet, leaving approximately 20 to 30 μl.

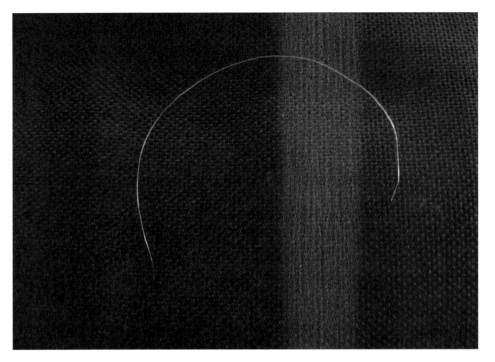

FIGURE 3.5   Hair simulated evidence sample.

3. Transfer filter paper to a sterile 1.5-mL microcentrifuge tube containing a spin basket or blood pellet to a sterile 1.5-mL microcentrifuge tube.
4. Add 1 mL of 1X TE extraction buffer (or other desired extraction buffer) from the first step of the desired extraction procedure.

Part G: Liquid Sample (e.g., Buccal Cells)

1. Swab the inside of the cheek using two swabs (or toothbrush), rotating them during collection; allow to air dry prior to storage in plastic cover, or collect wet saliva on the marked area of filter paper or FTA card (Figure 3.6) and allow to dry.
2. Cut swabs into at least three equal pieces using sterile scissors.
3. Add pieces of two swabs (one each of first and second swab) to a sterile tube containing 1 mL of 1X TE extraction buffer (from the first step of the desired extraction procedure) containing a spin basket. Mix thoroughly.

Part H: Swab Sample (e.g., Blood, Saliva, Semen, Fingerprint, Dress Shirt, Hat Head Band, Soda Can)

1. Lightly moisten a sterile swab (Figure 3.7) with sterile water, rotate/rub over the stain, and allow it to air dry before storing. Continue to swab a second time with a dry swab.

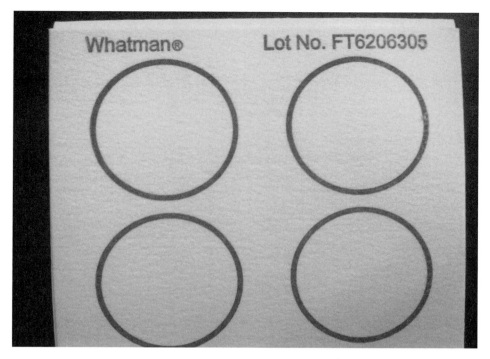

FIGURE 3.6 FTA card (Whatman).

2. Cut swabs into at least three equal pieces.
3. Add pieces of two swabs (one each of first and second swab) to a sterile tube containing 1X TE extraction buffer (from the first step of the desired extraction procedure) containing a spin basket. Mix thoroughly.

FIGURE 3.7 Swab collection device with 50 μL of blood.

Part I: Fingernail Clippings or Scrapings

1. Use clean clippers to clip nails into clean paper, double swab, or scrape undersides of fingernail into clean paper.
2. Store in a druggist's fold.
3. Continue to extract where the sample will be added to 1X TE extraction buffer (from the first step of the desired extraction procedure). Mix thoroughly.

Part J: Victim Sexual Assault Kit

Collect the swab, microscope slide, combing, and other samples using prescribed directions and contents of kit. Proceed as explained earlier for swabs.

Part K: Contaminated Item with Spatter/Pattern

1. If a bloodstain (Figure 3.8) or other spatter pattern is present, collect the entire item.
2. Cut stain from the item for analysis.
3. Cut stain into at least three equal pieces and add to sterile 1.5-mL microcentrifuge tubes containing spin baskets.
4. Cut a 1-cm section of the unaffected material adjacent to the stain, and place it in a separate tube to be utilized as a substrate sample (substrate control).
5. Add 1 mL of 1X TE extraction buffer (or other desired extraction buffer) from the first step of the desired extraction procedure. Mix thoroughly.

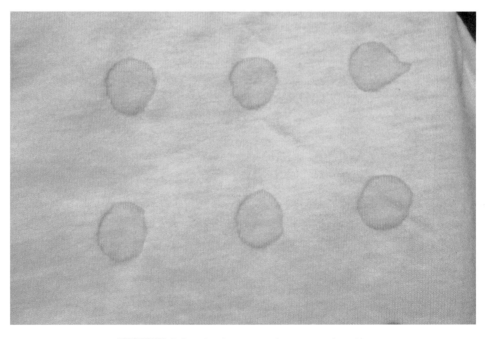

FIGURE 3.8   Bloodstain on white cotton (50 μL).

QUESTIONS

Which DNA sources will yield the most DNA? Which DNA sources will yield the least DNA? Which DNA sources are most important for initial analysis by the crime lab and why?

References

Bond, J.W., Hammond, C., 2008. The Value of DNA Material Recovered from Crime Scenes. J. Forensic Sci. 53, 797–801.

Drobnic, K., 2003. Analysis of DNA evidence recovered from epithelial cells in penile swabs. Croatian Medical Journal 44, 350–354.

Goray, M., van Oorschot, R.A., Mitchell, J.R., 2012. DNA transfer within forensic exhibit packaging: potential for DNA loss and relocation. Forensic Sci. Int. Genet. 6, 158–166.

Lorente, M., Entrala, C., Lorente, J.A., Alvarez, J.C., Villanueva, E., Budowle, B., 1998. Dandruff as a potential source of DNA in forensic casework. J. Forensic Sci. 43, 901–902.

President's DNA Initiative, http://www.dna.gov/training/, accessed 11/21/11.

Rutty, G.N., Hopwood, A., Tucker, V., 2003. The effectiveness of protective clothing in the reduction of potential DNA contamination of the scene of crime. Int. J. Legal Med. 117, 170–174.

4

DNA Extraction

OBJECTIVE

To learn how to perform DNA extractions using the Chelex, phenol-chloroform-isoamyl alcohol (PCIA), and commercial kit methods.

SAFETY

The PCIA reagent should remain in the hood and should be used with care; use the designated waste container in the hood for phenol-chloroform waste. Phenol is extremely corrosive to skin and eyes and can cause severe burns, so use double gloves. Chloroform is a carcinogen and is toxic when inhaled, absorbed through the skin, or ingested. Handle all kits and kit materials with gloves to protect the contents from contamination from skin cells and nucleases.

MATERIALS

1. Nuclease-free water
2. Micropipettes (0.5 to 10 μL, 10 to 20 μL, 10 to 200 μL, and 100 to 1000 μL) and autoclaved tips
3. Waterbath or heating block set to 56° C
4. 1.5-mL sterile microcentrifuge tubes
5. 0.9% saline (w/v) solution (prepared from food-grade table salt)
6. 50-mL sterile centrifuge tubes
7. Refrigerated centrifuge with rotor for 50-mL tubes
8. Microcentrifuges for 1.5-mL tubes
9. Vortexers
10. Gloves
11. Refrigerator or freezer to store extracted DNA
12. TE Extraction buffer (10 mM Tris-0.100 mM EDTA, pH 8)
13. 5% Chelex solution
14. 10 mg/mL Proteinase K
15. Organic stain extraction buffer (0.2 M NaCl, 0.02 M Tris (pH 8), 0.02 M EDTA, 5% SDS)

16. Phenol-chloroform-isoamyl alcohol (pH 7.8 to 8.2), water-saturated
17. 1 M dithiothreitol
18. Amicon or Microcon YM-100 filters or Vivacon filters
19. DNA IQ System by Promega and magnetic stand
20. QiaAMP by Qiagen
21. PrepFiler by Applied Biosystems

(Notes: Item 13 is for Chelex only, items 15 to 18 are for PCIA only, item 19 is for DNAIQ only, item 20 is for QiaAMP only, and item 21 is for PrepFiler only.)

RECIPES FOR BUFFER AND SOLUTION PREPARATION

1 M Tris (pH 8)

1. Dissolve 121.1 g of Tris base in 800 mL of ultra pure water.
2. Adjust pH to 8 ($\pm$ 0.2) with concentrated HCl.
3. Bring final volume up to 1 L with ultra pure water.
4. Autoclave.

0.5 M EDTA

1. Slowly add 186.1 g of $Na_2EDTA*2H_2O$ to 800 mL of ultra pure water.
2. Stir with a magnetic stir bar.
3. Adjust pH to 8 ($\pm$ 0.2) by adding NaOH pellets.
4. Bring final volume up to 1 L with ultra pure water.
5. Autoclave.

20% SDS

1. Add 200 g of SDS to 800 mL of ultra pure water (slow to dissolve; may need heat).
2. Bring final volume up to 1 L with ultra pure water.
3. Mix well.

TE Buffer

1. Mix together the following items:
    10 mL of 1 M Tris-HCl (pH 8)
    0.2 mL of 0.5 M EDTA
    990 mL of ultra pure water
2. Autoclave.

Organic Stain Extraction Buffer

1. Dissolve 5.84 g of NaCl in 500 mL of ultra pure water.
2. Add 10 mL of 1 M Tris (pH 8), 20 mL 0.5 M EDTA, and 100 mL 20% SDS.

3. Titrate to pH 8 ($\pm$ 0.2) with HCl.
4. Bring final volume up to 1 L with ultra pure water.
5. Autoclave.

Proteinase K

1. Dissolve 100 mg Proteinase K in 10 mL of ultra pure water.
2. Aliquot at 0.5 mL each and store in freezer.

Dithiothreitol (DTT)

1. Add 1.5424 g of DTT to 10 mL of ultra pure water
2. Mix well and aliquot 1 mL each; store in freezer.

TAE Buffer

1. Add 20 mL of 50X TAE to 980 mL of deionized water (total volume 1 L) OR Add 242 g Tris base (FW 121.14) to approximately 750 mL of deionized water. Add 57.1 mL of glacial acetic acid, 100 mL of 0.5 M EDTA, and bring final volume to 1 L. The pH should be approximately 8.5.

BACKGROUND

Deoxyribonucleic acid (DNA) extraction is the process by which DNA is separated from proteins, membranes, and other cellular material contained in the cell from which it is recovered. This extraction can be one of the most labor-intensive parts of DNA analysis. Extraction methods may require an overnight incubation, may be a protocol that can be completed in minutes or a couple of hours, or may be a recent procedure that employs reagents for which this step can be skipped completely.

The DNA extraction process requires careful handling of biological material to prevent sample contamination and crossover. Tubes should be carefully labeled, especially when transfers are required. Robots may be employed to extract reference samples and some evidence samples, but other evidentiary samples may require the direct attention of a DNA analyst.

The simplest cells, such as bacteria cells, are prokaryotes. These prokaryotes comprise a lipid bilayer outer membrane and a cytoplasm containing a circular chromosome, proteins inorganic salts and metal ions, sugar molecules, and other elements of cell machinery. Humans, animals, and plants are composed of eukaryotic cells; these cells also have a lipid bilayer outer membrane and cytoplasm containing proteins, sugars, lipids, and inorganic ions of various types and function. However, eukaryotic cells also contain other membrane-enclosed compartments called organelles. The nucleus of a cell is an organelle that houses 46 chromosomes, and the mitochondria each house a circular DNA chromosome, all of which direct the production of proteins. The mitochondrial chromosome is

16,569 bp. The mitochondrial proteins are used as some of the metabolic machinery for digestion of sugars and fats and production of most of the energy for the cell. Other organelles are involved in the synthesis and modification of proteins, sugars, and fats, and other molecules used by the cell or in its signaling activities. The genes that code for heritable traits—including height, blood type, hair color, eye color, skin color, and temperament—are found on chromosomes in the nucleus of the eukaryotic cell. In plants, the additional chloroplast chromosome contains genes for photosynthesis. Forensic scientists are largely unconcerned with the genes that code for proteins or regulatory elements of the DNA; however, forensic scientists are interested in isolating DNA from the nuclear, mitochondria and chloroplast (if present) chromosomes to evaluate the sequences, base and repeat polymorphisms, including SNPs and STRs, respectively, that have proved useful for linking a suspect to a crime scene.

DNA is highly negatively charged because of its phosphate groups. It is stabilized by magnesium in the cell when unwound. Nuclear DNA consists of 3.2 billion bases in humans. It is organized into chromosomes in part by coiling around positively charged proteins called histones to form nucleosomes. Magnesium is also integral to the function of proteases, enzyme proteins that cut up DNA.

Because of the lipid structure of the cell (and nuclear) membrane(s), presence of proteases and magnesium, and coiling of DNA around histones, many of the available DNA extraction procedures have common elements. Indeed, the extraction of DNA generally follows three basic steps:

1. Lyse (break open) the cells.
2. Separate the DNA from the other cell components.
3. Isolate the DNA.

The cell membrane is disrupted by any of the following methods: using heat to increase fluidity, dithiothreitol (DTT) to reduce disulfide bonds, or a detergent, such as sodium dodecyl sulfate (SDS), to disrupt the membrane. Proteins, including nucleases, are inactivated by heat denaturation or by digestive enzymes, including proteinase K, to cut them up. The temperature must be kept below $60°$ C and the period must be kept sufficiently short (15 to 20 minutes) if nondegraded, high-molecular-weight DNA is required. Either the magnesium needed for nuclease activity or DNA is immobilized on a solid phase and eluted by buffer/salt. If the DNA remains in the aqueous phase, it is separated from the other cellular materials including proteins and lipids by centrifuging the latter to the bottom of the tube or partitioning them in organic solvents.

There are four commonly used extraction procedures for DNA extraction (Hoff-Olsen et al., 1999):

1. Organic (variations of phenol/chloroform): use of a multistep liquid chemical process that is labor intensive but produces a high yield and very-clean double-stranded extracted DNA sample
2. Inorganic Chelex or silica methods: simple and cheap one-tube extraction process in which Mg^{2+} binds to resin beads and yields a single-stranded DNA product
3. Solid phase extraction methods (e.g., Promega's DNA IQ (Eminovic et al., 2005, DNA IQ manual), Applied Biosystems' PrepFiler (Brevnov et al., 2009, PrepFiler manual), and

Qiagen's QIAamp kits (Castella et al., 2006, Greenspoon et al., 1998)): simple extraction process in which the DNA binds to paramagnetic or silica beads

4. Differential extraction: a multistep process used to separate sperm from other cells using DTT; used for analyzing biological evidence from sexual assault cases (Drobnic, 2003)

For organic extraction, the cells must be lysed using gentle heat to disrupt the membranes with the assistance of SDS and DTT. Proteinase K must be added to digest the proteins, including nucleases. In the second step of the PCIA extraction, the DNA must be separated from the other cellular components using the phenol-chloroform-isoamyl alcohol (PCIA) reagent. When the PCIA is added, two phases appear: the organic (bottom) and the aqueous (DNA-containing) phase (top) after centrifugation; the DNA-containing portion is removed by pipetting off the top layer. DNA is more soluble in the aqueous state compared to the organic phases. It is important to remove all of the organic extraction agents because phenol will degrade DNA. Phenol has a density of approximately 1 g/mL and can easily invert with the aqueous phase. Lastly, in the third step, the DNA can be filtered and concentrated by centrifugation through a Microcon, Centricon, or Vivacon 100 filter in which the DNA is captured on the membrane while all other aqueous components including protein fragments, residual organics, and potential PCR inhibitors are allowed to pass through the membrane (100 kilodalton cut-off) (Moore, 1998). The DNA is eluted by inverting the filter and centrifuging using a desired volume of elution buffer or nuclease-free water (20 to 50 µL). Using the filter, the DNA is concentrated in a small volume for subsequent amplification steps. This method yields a double-stranded DNA extract and the highest yield but requires multiple tube changes which increase the possibility of contamination error.

The basic Chelex procedure (Walsh et al., 1991) consists of an overnight incubation to disrupt the cell membrane and allow the 5% Chelex solution (styrene divinylbenzene copolymers containing paired iminodiacetate ions), which has a high affinity for and binds polyvalent metal ions such as magnesium, and Proteinase K to digest the nucleases, and boiling the sample in a 5% Chelex solution to burst the remaining membranes, denature the proteins and denature the DNA in the cells. The result is a denatured sample of single-stranded DNA that remains in the supernatant after centrifugation. The single-stranded DNA is then ready to be concentrated, and if necessary it can be quantified and amplified. Because this is a single-tube reaction, there is less potential for contamination and mis-pipetting. Silica and glass beads can be substituted for Chelex in this type of extraction procedure (Dederich et al., 2002).

The DNA IQ System by Promega employs silica-coated paramagnetic beads, which bind the DNA (up to 100 ng sample). The magnetic beads are drawn to the bottom of the tube using a magnetic holder, and the other cellular components are washed away. The process may be automated using robotic systems and is ideal for consistent-concentration samples including paternity and reference samples. The kit contains the proprietary resin and specialized lysis, elution, and wash buffers for the procedure (Eminovic, et al., 2005, DNA IQ manual).

The Qiagen QIAamp DNA Micro Kit employs silica-based bead extraction method using a spin column. The silica beads bind the DNA under high chaotropic salt conditions. The kit contains the spin columns and sample and lysis buffer, none of which contain hazardous chemicals. The cell lysate is combined with alcohol and placed into the spin column, which is inserted into a tube. Proteins and divalent cations, positively charged ions, including magnesium, are removed using multiple buffer washes (specific formulations are available for different cell/sample types) and centrifugation steps. Pure DNA is eluted from the membrane using the desired quantity of sterile water or Tris-EDTA buffer. The procedure can be automated (Castella et al., 2006, Greenspoon et al., 1998).

FTA filter paper cards use technology licensed by Whatman from Flinders University. The technique was originally developed by Burgoyne and Fowler at Flinders University in Australia in the 1980s. The nucleic acid is protected from degradation by nucleases from reagents contained in the filter paper including a weak base, a chelating agent, an anionic surfactant or detergent, and uric acid (or a urate salt) to a cellulose-based matrix (filter paper). A sample containing DNA can then be applied to the treated filter paper for preservation and long--term storage. The cards are widely used because they are small, can be stored at room temperature, and ship easily (Thacker et al., 2000).

In this experiment DNA will be extracted from questioned and known samples containing biological material using one or more of the following methods in one to two sessions.

PROCEDURE

Week 1

Isolation of Cells from Simulated DNA Evidence Samples (Continued from Chapter 3)

1. For Chelex extraction, proceed directly to Chelex procedure. The cells are in 1 mL of 1X TE buffer. For other methods, centrifuge the 1X TE liquid to collect the cells into a pellet. Remove the swabs and spin basket from the liquid, and proceed with extraction process.
2. For PCIA, QiaAMP, DNAIQ, and PrepFiler, centrifuge the tubes at maximum speed for 2 minutes. Remove supernatant, being careful not to disturb pellet, leaving approximately 20 to 30 µL.
3. Reconstitute the pellet and proceed with the desired extraction process.

Buccal Cell Collection (only necessary if DNA-containing samples were not isolated in the sample collection lab; yields sufficient DNA to be detectable before PCR amplification on an agarose gel)

1. Obtain approximately 15 mL of the 0.9% w/v saline solution to gargle with to obtain cheek cells. Swish hard and chew on your cheeks (but don't draw blood). Collect the solution in a sterile 50-mL centrifuge tube. Repeat two to three times to yield approximately 45 mL.
2. Centrifuge at 4000 rpm for 10 minutes at 4° C to pellet the cheek cells. Balance your saline-cell mixture with that of a similar volume from a classmate or with a tube

containing an equal mass of water. After centrifugation, immediately remove the centrifuge tubes. Carefully pipette or decant off the supernatant, so as to not disturb the cheek cell pellet.

3. Transfer the approximately 1 mL pellet equally to two sterile 1.5-mL microcentrifuge tubes.
4. Centrifuge in a microcentrifuge for 30 to 60 seconds. Again, carefully pipette off the supernatant, so as to not disturb the cheek cell pellet. The pellet is now ready for extraction. This sample will serve as a positive control, as extraction should yield sufficient cells for further labs.

Chelex (Inorganic) Extraction

1. Add 1 mL of TE Buffer to a 1.5-mL microcentrifuge tube containing half of the buccal cell pellet sample or other collected simulated evidence sample (if not already done).
2. Incubate at room temperature for 15 to 30 minutes. Mix occasionally by inversion or gentle vortexing.
3. Spin in a microcentrifuge for 2 to 3 minutes at 10,000 to 15,000x gravity. If substrate is present, use a spin basket to separate the substrate from the aqueous portion.
4. Carefully remove the supernatant (all but 20 to 30 μL, without disturbing pellet), and transfer to a sterile microcentrifuge tube. Discard the pellet.
5. Add 200 μL of 5% Chelex solution to the supernatant.
6. Add 2 μL of Proteinase K (10 mg/mL).
7. Incubate at 56° C for 6 to 8 hours or overnight. This incubation may be reduced to as little as 30 minutes with a possible reduction in yield but sufficient DNA for PCR and DNA typing.

Phenol-Chloroform-Isoamyl Alcohol (Organic) Extraction

1. If the sample is in 1X TE buffer, centrifuge to pellet cells and pipet off buffer. To cells, add the 300 μL Organic Stain Extraction Butter, 12 μL of 1 M DTT, and 4 μL of 10 mg/mL Proteinase K to a 1.5-mL microcentrifuge tube containing half of the buccal cell pellet sample or other collected simulated evidence sample.
2. Vortex the mixture in a microcentrifuge tube for 2 to 5 seconds.
3. Incubate at 56° F overnight.

Promega DNA IQ Extraction

1. If the sample is in 1X TE buffer, centrifuge to pellet cells and pipet off buffer. Obtain the DNA IQ kit (Figure 4.1). To each sample containing cells to be extracted, add 150 to 250 μL of prepared Lysis Buffer (and 1 μL of 1 M DTT for each 100 μL Lysis Buffer), mix by inversion several times, and incubate at 70° C for 30 minutes.
2. Transfer Lysis Buffer and sample into DNA IQ Spin Basket.
3. Centrifuge for 2 minutes at maximum speed in microcentrifuge.
4. Vortex resin for 10 seconds. Remove spin basket. Add 7 μL resuspended resin to sample.

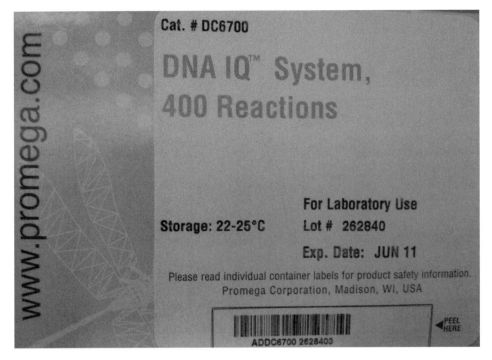

FIGURE 4.1   DNA IQ kit (Promega).

5. Vortex for 3 seconds at high speed, and incubate at room temperature for 5 minutes. Vortex once a minute for 3 seconds each time while incubating.

6. Vortex for 2 seconds, and place in magnetic stand (MagneSphere Technology Magnetic Separation Stand [two-position]). Separation should occur instantly, and the magnetic beads should be at the bottom of the tube.

7. Carefully discard solution without disturbing resin on side of tube (pipette).

8. Wash with 100 μL prepared Lysis Buffer and vortex 2 seconds at high speed.

9. Return the tube to the magnetic stand, and discard all Lysis Buffer.

10. Wash with 100 μL prepared wash buffer and vortex 2 seconds at high speed.

11. Return the tube to the magnetic stand, and discard all wash buffer.

12. Wash three times (repeat steps 9 to 10 twice more) with 1x wash buffer.

13. Air-dry at room temperature for 5 minutes (but not more than 20 minutes).

14. Add 25 to 100 μL of Elution Buffer, close lid, and vortex for 2 seconds. Incubate at 65° C for 5 minutes.

15. Remove tubes from heat, vortex for 2 seconds, and place on magnetic stand.

16. Remove DNA solution while still hot and transfer to a sterile 1.5-mL microcentrifuge tube.

17. DNA can be stored at 4° C for short-term storage or at −20° or −70° C for long-term storage.

PrepFiler

1. Obtain PrepFiler kit. For a swab sample, up to 40 μL of blood or saliva, or up to 25-mm^2 cutting/punch of blood (on FTA or fabric) or semen/saliva on fabric. If sample is in 1X TE buffer, centrifuge to pellet cells and pipette off buffer.

2. Prepare a water bath or thermal shaker at 70° C.

3. Place sample in a PrepFiler spin tube or 1.5-mL microcentrifuge tube, and add 300 μL PrepFiler lysis buffer and 3 μL 1 M DTT (5 μL for samples containing semen). The fluid should cover the substrate.

4. Mix by gentle vortexing for 5 seconds.

5. Spin in microcentrifuge briefly.

6. Incubate sample at 70° C and 900 rpm for 20 minutes for liquid body fluid samples (increase to 40 minutes for dried stains or samples on swabs and 90 minutes for neat semen samples).

7. For a sample with a swab or fabric (or other) substrate, centrifuge the microcentrifuge tube containing the sample for 2 seconds.

8. Insert a PrepFiler Filter Column into a new 1.5-mL PrepFiler Spin Tube, then carefully transfer the sample tube contents into the filter column, using a pipette to transfer the liquid and the pipette tip or sterile tweezers to transfer the substrate.

9. Close the cap on the filter column/spin tube and centrifuge at maximum speed (2 minutes at 12,000 to 14,000 rpm; 5 minutes at 3000 to 4000 rpm) until at least 180 μL sample lysate is collected in the spin tube.

10. Remove the PrepFiler Filter Column from the Spin Tube and dispose of it properly. Retain the sample in the Spin Tube. You may transfer sample lysate to a different sterile 1.5-mL microcentrifuge tube if the cap does not securely close.

11. Wait approximately 5 minutes or until sample lysate is at room temperature (do not chill).

12. Vortex the PrepFiler Magnetic Particles tube approximately 5 seconds to mix. Check that the particles are suspended and that no visible particles remain in the bottom of the tube. Centrifuge briefly. Continue to resuspend particles every 5 minutes when working with multiple samples.

13. Pipet 15 μL of magnetic particles into the tube containing the sample lysate. Close the tube and vortex for 10 seconds at 500 to 1200 rpm.

14. Centrifuge briefly, then add 180 μL of isopropanol to the sample lysate tube.

15. Close tube and vortex for 5 seconds at 500 to 1200 rpm.

16. Centrifuge briefly then put the sample lysate tube in a shaker or on a vortexer (with adaptor), then mix at room temperature at 1000 rpm for 10 minutes.

17. If magnetic particles are present on the sides of the sample DNA tube above the meniscus, invert the tube to resuspend the particles and vortex the sample lysate DNA tube at maximum speed (approximately 10,000 rpm) for 10 seconds, then centrifuge briefly.

18. Place the sample DNA tube in the magnetic stand and wait until the size of the pellet of magnetic particles on the back of the tube stops increasing (approximately 1 to 2 minutes).

19. Pipet off the visible liquid phase while the sample DNA tube remains in the magnetic stand. Do not aspirate.

20. Add prepared wash buffer to the sample DNA tubes using a sequence of 600 µL wash buffer A (first wash), 300 µL wash buffer A (second wash), 300 µL wash buffer B (third wash). Cap the sample DNA tube and from the magnetic stand to vortex at maximum speed (approximately 10,000 rpm) until the magnetic pellet is dispersed and no longer visible, centrifuge briefly, and replace the sample DNA tube in the magnetic stand for 30 to 60 seconds before removing the wash liquid. Repeat this step four more times.

21. Open the DNA sample tube in the magnetic stand and air-dry the DNA for 7 to 10 minutes at <25° C (air-drying for more than 10 minutes may reduce DNA yield). If the room temperature is >25° C, reduce drying time to 5 minutes.

22. Elute DNA by adding 50 µL of PrepFiler Elution Buffer or 1X TE or 0.1X TE to the sample DNA tube. Do not use water.

23. Cap sample tube and vortex at maximum speed (approximately 10,000 rpm) for 5 seconds or until the magnetic particle is dispersed.

24. Place the tube in a thermal shaker temperature at 70° C for 5 minutes at 900 rpm. If a shaker is not available, heat in a heating block and vortex and centrifuge every 2 to 3 minutes.

25. Put the tube in the magnetic stand until the pellet forms. Pipet the supernatant (which contains the isolated genomic DNA) onto a sterile 1.5-mL microcentrifuge tube for storage. Store the DNA at 4° C for up to 1 week, or at −20° C for long-term storage.

Week 2

Chelex (Inorganic Extraction)

8. Remove the sample from the heating block. Vortex the sample at high speed for 5 to 10 seconds.

9. Spin in a microcentrifuge briefly at 10,000 to 15,000 x g or high speed.

10. Incubate in a boiling water bath or heating block (>100° C) for 8 minutes.

11. Remove and vortex at high speed for 5 to 10 seconds.

12. Spin in microcentrifuge for 2 to 3 minutes at 10,000 to 15,000 x g or high speed.

13. Pipet the supernatant containing the DNA into a new sterile 1.5-mL microcentrifuge tube. Discard the tube containing the pellet of Chelex beads and cell debris.

14. The samples are now ready for the quantification and amplification processes. Store the samples at 4° C for short term or freeze for long term (−20° C or −80° C) in Teflon-seal tubes. To reuse, thaw samples and repeat steps 10 and 11.

Phenol-Chloroform-Isoamyl Alcohol (Organic) Extraction

4. Remove the sample from the heating block. Centrifuge for 2 to 5 seconds.

5. Add 300 µL of the phenol-chloroform isoamyl alcohol (PCIA) reagent (Figure 4.2) in the fume hood.

6. Vortex for 2 to 5 seconds (briefly) or until the formation of a milky emulsion.

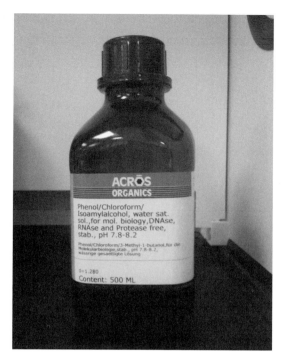

FIGURE 4.2   Commercially available PCIA reagent (pH 7.8 to 8.2).

7. Centrifuge for 3 minutes in a microcentrifuge at maximum speed to separate the layers.
8. Assemble a Microcon (Figure 4.3), Vivacon, or Centricon-100 centrifugation unit by placing the filter into their microcentrifuge tubes with long armed caps.
9. To the top of the concentrator (Figure 4.4), add 100 µL 1X TE Buffer. Transfer the aqueous phase (top) from the tube in step 7 to the top of the concentrator. Avoid pipetting organic solvent from the tube into the concentrator. If large concentrator units are used, adjust the volumes accordingly.
10. Centrifuge for 10 minutes at 2500 x g. If the entire sample did not fit, spin briefly and discard eluent and add the remainder.
11. Add 100 µL of 1X TE Buffer to top of filter to wash.
12. Centrifuge for 10 minutes at 2500 x g. Repeat steps 11 and 12 as desired.
13. Add 20 to 50 µL of 1X TE Buffer (depending on desired concentration) to the top of the concentrator and flip it (top down) (Figure 4.5) onto another sterile 1.5-mL microcentrifuge tube.
14. Centrifuge the tube containing the DNA for 5 minutes.
15. Discard the concentrator and close top of the microcentrifuge tube to secure the DNA sample. Label the tube if you have not already done so.

FIGURE 4.3   Microcon Centrifugal Filter Device.

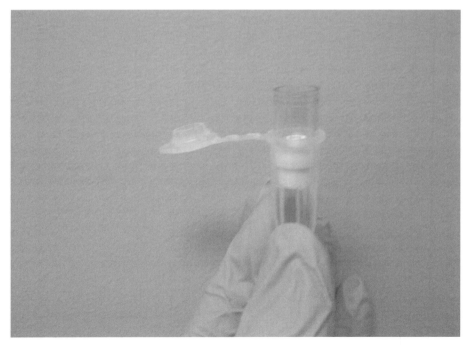

FIGURE 4.4   Assembling Microcon Centrifugal Filter Device to filter and concentrate the sample.

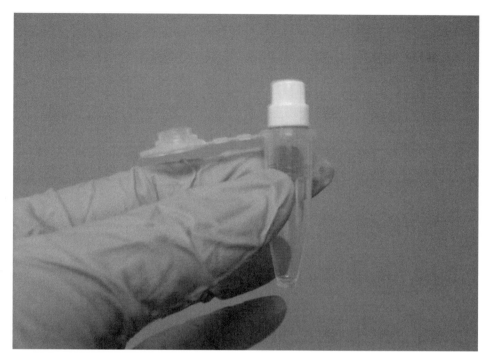

FIGURE 4.5   Assembling Microcon Centrifugal Filter Device to elute the sample.

16. The samples are now ready for the quantification and amplification processes. Store the samples at 4° C for short term or freeze for long term (−20° C or −80° C) in Teflon-seal tubes.

QUESTION

Note the changes and appearance of the solution after each step and each reagent addition. If possible, describe the appearance of the crude DNA.

References

Brevnov, M.G., Pawar, H.S., Mundt, J., Calandro, L.M., Furtado, M.R., Shewale, J.G., 2009. Developmental validation of the PrepFiler Forensic DNA Extraction Kit for extraction of genomic DNA from biological samples. J. Forensic Sci. 54, 599−607.

Castella, V., Dimo-Simonin, N., Brandt-Casadevall, C., Mangin, P., 2006. Forensic evaluation of the QIAshredder/ QIAamp DNA extraction procedure. Forensic Sci. Int. 156, 70−73.

Dederich, D.A., Okwuonu, G., Garner, T., Denn, A., Sutton, A., Escotto, M., et al., 2002. Glass bead purification of plasmid template DNA for high throughput sequencing of mammalian genomes. 30, 1−5.

Drobnic, K., 2003. Analysis of DNA evidence recovered from epithelial cells in penile swabs. Croatian Medical Journal 44, 350e–354e.

DNA IQ manual, Promega, Madison, WI.

Eminovic, I., Karamehić, J., Gavrankapetanović, F., Heljić, B., 2005. A simple method of DNA extraction in solving difficult criminal cases. Med. Arh. 59, 57–58.

Greenspoon, S.A., Scarpetta, M.A., Drayton, M.L., Turek, S.A., 1998. QIAamp spin columns as a method of DNA isolation for forensic casework. J. Forensic Sci. 43, 1024–1030.

Hoff-Olsen, P., Mevag, B., Staalstrom, E., Hovde, B., Egeland, T., Olaisen, B., 1999. Extraction of DNA from decomposed human tissue: An evaluation of five extraction methods for short tandem repeat typing. Forensic Sci. Int. 105, 171–183.

Kolman, C.J., Tuross, N., 2000. Ancient DNA analysis of human populations. Am. J. Phys. Anthropol. 111, 5–23.

Moore, D., 1998. Purification and concentration of DNA from aqueous solutions.. In: Current protocols in pharmacology. John Wiley & Sons, Inc. A.3C.1-A.3C.7.

PrepFiler manual, Applied Biosystems.

Thacker, C., Phillips, C.P., Syndercombe-Court, D., 2000. Use of FTA cards in small volume PCR reactions. Progress in Forensic Genetics 8, 473–475.

Walsh, P.S., Mitzger, D.A., Higuchi, R., 1991. Chelex-100 as a medium for simple extraction of DNA for PCR-based typing from forensic material. BioTechniques 10, 506–513.

Determination of Quality and Quantity of DNA Using Agarose Gel Electrophoresis

OBJECTIVE

To learn how to run agarose gel electrophoresis and interpret the gel to assay the size and quality of DNA extracted from simulated crime scene evidence.

SAFETY

Wear gloves at all times. SYBR Green I is an intercalating agent. The gels are run at high voltage and pose an electrocution hazard. Set the digital camera to UV-Vis Safety. Record gel pictures with the illuminator door in the closed position.

MATERIALS

1. Gloves
2. Extracted DNA
3. Micropipettes (0.5 to 10 μL, 10 to 20 μL, 10 to 200 μL, and 100 to 1000 μL) and autoclaved tips
4. Waterbath or heating block set to 100° C
5. Agarose (molecular biology grade)
6. Horizontal gel electrophoresis assembly with 14-well comb
7. Power supply
8. 10 kb to 100 bp range or similar DNA ladder
9. 6X Loading dye
10. 50X TAE buffer, diluted to 1X TAE
11. 100X SYBR Green I dye
12. 1.5-mL microcentrifuge tubes

13. Microcentrifuges
14. Vortex machines
15. Digital photodocumentation system
16. Microwave or hot plate

BACKGROUND

Electrophoresis is the process of separating molecules based on their polarity. In the mid-1930s, Arne Tiselius observed that if ions of similar charge are placed in solution between two oppositely charged electrodes, the smaller ions of the same charge move faster toward the electrode of opposite charge than the larger-sized ions. If ions of varying charge are placed in a solution, the more highly charged ions migrate faster than the lower charged ions. Because DNA has a negative charge, it will migrate through the gel toward the cathode or positive electrode. He was awarded the 1948 Nobel Prize in Chemistry for his work.

Forensic scientists can use agarose gel electrophoresis to determine the quality and quantity of DNA recovered from a sample. Agarose is a sulfonated polysaccharide, or carbohydrate, whose molecules consist of a number of sugar molecules bonded together. Isolated from marine algae, it is used in the laboratory as a matrix support for gels in electrophoresis. At room temperature, it is an insoluble powder that can be dissolved by heating it on a hot plate or in a microwave. When boiled in a running buffer, the solution will be viscous and transparent, allowing it to be poured into a mold that will harden to a gel with the consistency of a gelatinized substance. As it solidifies, the gel's color will change to a translucent, faint blue. A well comb is used while the gel is still cooling to produce cutouts in the gel, termed wells, where samples can be loaded. Run times vary depending on the percentage of agarose used but are typically 30 to 60 minutes for low-percentage gels. The voltage must not exceed 15 volts per centimeter, as the polymerization is reversible and the gel can melt.

Agarose gel electrophoresis is generally used to evaluate extracted high-quality, standard-reference genomic DNA from cell lines and other samples. A tight band indicates that the DNA is intact, and a smear indicates that the DNA is degraded (Till et al., 2006). Additionally, gel electrophoresis allows for the determination of DNA quantity in a sample using quantification standards. A faint band indicates a low quantity of DNA, and an intense band indicates a high quantity of DNA.

Because gel electrophoresis is a relative technique, the migration of the DNA extract must always be compared to a size standard starting at the same origin on the same gel. Size standard ladders are used to estimate the size of the extracted genomic DNA. A calibration curve (Figure 5.1) can be produced by plotting the log of the molecular weight of the DNA ladder standards (in base pairs) against the distance each has migrated from the well (Table 5.1). The distance that the unknown piece of DNA has migrated can be used to determine its molecular weight graphically using a regression line and slope from the line equation.

As an alternative to agarose gels, polyacrylamide gels can be used to separate small fragments of DNA or ribonucleic acid (RNA) (smaller than 100 bp); however, modern techniques using 3% agarose gels can show up to one base pair size difference with an extremely low conductivity medium.

FIGURE 5.1 Plot of log base pairs of DNA size standard ladder fragments versus distance traveled by the DNA fragments in millimeters.

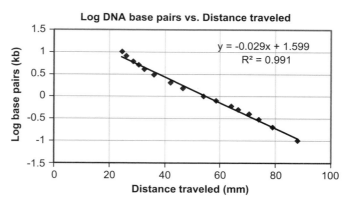

Because of electrophoresis, the shorter DNA fragments will migrate faster (and farther) than longer DNA fragments in a given run time. The shorter fragments can form a ball and tumble through the matrix according to the Ogsten sieving theory or linearly snake through the matrix according to the reptation theory. The distance the DNA will migrate from the well is related to the length (in base pairs) of the DNA, the structure of the DNA

TABLE 5.1 Sample Data for Figure 5.1. Size of DNA Fragments in Base Pairs and Distance Migrated on Gel.

Size (bp)	Size (kb)	Distance (cm)	Log kb	Distance (mm)
10000	10	2.45	1	24.5
8000	8	2.60	0.90309	26.0
6000	6	2.85	0.778151	28.5
5000	5	3.05	0.69897	30.5
4000	4	3.25	0.60206	32.5
3000	3	3.60	0.477121	36.0
2000	2	4.20	0.30103	42.0
1500	1.5	4.65	0.176091	46.5
1000	1	5.40	0	54
800	0.8	5.85	-0.09691	58.5
600	0.6	6.40	-0.22185	64.0
500	0.5	6.65	-0.30103	66.5
400	0.4	7.05	-0.39794	70.5
300	0.3	7.40	-0.52288	74.0
200	0.2	7.90	-0.69897	79.0
100	0.1	8.80	-1	88.0

(supercoiled, relaxed, circular, or linear), and the degree of complexing of the agarose matrix (concentration). High-quality genomic DNA will migrate little, if any, from the loading well; it will remain at the top of the agarose gel and migrate as an intact band. As few as 20 nanograms of DNA can be visualized using modern fluorescent dyes. It is easy to visualize, analyze, and even to reextract the DNA sample (the sample is not degraded and can be recovered from the gel using a razor blade) as needed.

A loading dye containing bromophenol blue, xylene cyanol, and orange G is often used to visualize the DNA sample loaded into the gel. In a 1% agarose gel, xylene cyanol migrates at a rate equivalent to 4000 bp DNA, bromophenol blue migrates at a rate equivalent to 200 to 400 bp DNA, and orange G migrates at approximately 50 to 100 bp. The loading dye typically contains glycerol or sucrose to increase the density of the DNA solution so that it settles to the bottom of the well and is not lost while loading.

In this experiment, the size and concentration of the extracted DNA will be estimated using agarose gel electrophoresis and the DNA size and concentration ladder.

PROCEDURE

1. To prepare a 1% agarose gel, combine 0.5 g of agarose and 50 mL of 1x TAE Buffer in a 250-mL Erlenmeyer.
2. Heat in the microwave for 40 seconds or until it boils. Remove from microwave. (Plastic wrap may be used to cap the flask to reduce buffer evaporation.) The agarose should be fully dissolved, and the solution should appear viscous and clear.
3. Cool the agarose-TAE solution to the touch, and pour it into the gel box positioned so that a rectangular box with four sides is created. Add a 14-well comb.
4. Once the gel is set (translucent blue color), rotate the gel so that the wells are positioned nearest the negative electrode (black).
5. Fill gel box with 1x TAE buffer to just cover the gel.
6. Carefully remove the comb from the gel.
7. To prepare samples for gel, add 1 μL of the 6X loading dye and 1 μL 100x SYBR Green I to 10 μL of each sample in microcentrifuge tubes.
8. Load the 12 μL each to separate wells carefully by pipetting to the middle lanes. Be careful not to puncture the bottom of the gel. Do not release the plunger until the pipette is removed from the gel so as to not draw up the samples from the well.
9. To prepare ladder for gel, combine 8 μL of the ladder (or the manufacturer recommended amount) with 1 μL of 100x SYBR Green I dye and 1 μL of the 6X loading dye.
10. Load ladder to two lanes on opposite sides of the gel.
11. Place the safety cover on the gel box to run the sample toward the positive (red) cathode. Connect and turn on the power supply.
12. Run the gel for 30 to 60 minutes at no more than 15 V/cm (or 150 V for 10 cm gel or shorter times for minigels). Agarose will melt if the voltage and heat are too high.
13. Turn off the power supply and unplug it. Remove the gel safety cover.
14. While wearing gloves, remove the gel from the casting plate, place it on a UV light box, and take a picture using a digital photodocumentation system employing a SYBR filter to detect DNA bands fluorescing green.

QUESTIONS

1. Sketch the gel or include a picture in your report.
2. Measure the migration distance of the DNA in the ladder, and compute the log of the base pairs for the ladder provided. Prepare a table.
3. Plot a graph of log base pairs versus distance traveled. Use the slope of the line to determine the size of the extracted DNA with its measured migration distance. What was the expected and observed size for the genomic DNA? Do the results differ for different extraction methods?

Reference

Till, B.J., Zerr, T., Comai, L., Henikoff, S., 2006. A protocol for TILLING and Ecotilling in plants and animals. Nature Protocols 1, 2465–2477.

Determination of DNA Quality and Quantity Using UV-Vis Spectroscopy

OBJECTIVE

To learn how to determine the concentration of a DNA sample by analyzing 260 nm and 280 nm absorbance values recorded using a UV-Vis spectrophotometer.

SAFETY

Do not look into the ultraviolet (UV) light source. Handle samples with gloves to avoid residual phenol depositing on the skin.

MATERIALS

1. Gloves
2. UV-Vis spectrophotometer
3. 1- or 3-mL quartz cuvette, 1-cm path length
4. Standard DNA for calibration curve (optional)
5. Extracted DNA
6. Micropipettes (0.5 to 10 μL, 10 to 20 μL, 10 to 200 μL, and 100 to 1000 μL) and tips

BACKGROUND

The use of ultraviolet (UV) or visible (vis) light to quantitate materials that absorb in the UV-vis regions of the electromagnetic spectrum is well documented. Both proteins and DNA have substantial absorbance in the UV region. The technique is nondestructive, and labs are commonly equipped with this instrumentation. It can also be used to determine the purity of protein or DNA.

Light consists of packets of energy or radiation called photons. The energy of the photons varies with the wavelength of the light; light of a long wavelength contains less energy than light of a shorter wavelength. Photons in ultraviolet light, shorter wavelengths, have a higher energy than photons contained in visible light, or longer wavelengths. Within visible light, red light has a longer wavelength and a lower energy than blue light. Compounds that absorb ultraviolet and visible light experience the excitation of certain electrons within the molecule from their ground energy state to an excited energy state. Absorbance is structure-dependent and reproducible.

Light that is not absorbed is transmitted (a solution irradiated with white light with a substance that absorbs red and yellow light transmits blue light which can be detected and recorded).

The advantages to UV-vis spectroscopy include that instruments are widely available, consumables and specialized reagents are not required, and the method is quick. Disadvantages include that the method is not specific to human DNA and that many substances including contaminating phenol reagent from extraction procedures absorb in the same region as DNA/RNA.

Other methods can also be used to quantitate DNA, including fluorescence-based techniques. Although UV spectroscopy can be used to quantitate samples of micromolar concentrations or greater, the fluorescence-based methods are more sensitive and can quantify samples of nanomolar concentration or greater. These alternate methods include the slot-blot technique, PicoGreen microtiter plate assay, AluQuant human DNA quantification system, real-time PCR, and fluorescence DNA quantification with the Hoechst reagent. Another drawback to the UV method is that it is not specific to human DNA. The absorbance will reflect the presence of contaminating phenol reagent, proteins, nonhuman DNA, and RNA. However, this method has the advantage of not requiring consumables and reagents and giving a rapid quantitation determination within the usable range. The UV spectrometer will allow for the determination of DNA concentrations of as little as a microgram (Holden et al., 2009, Gallagher 2011, Labarca and Paigen, 1980).

Spectrophotometry is based on the interaction of a substance with incoming radiation. Many molecular and atomic species that absorb in the UV region have characteristic absorption spectra. An absorption spectrum is obtained by plotting absorbance (A) as a function of wavelength (λ). The spectrum is a chemical fingerprint that can be qualitatively used to identify DNA present in a sample based on an absorption maximum at or near 260 nm depending on the composition of the DNA sample. Absorption spectroscopy can also be used to quantify the sample. If radiation of one wavelength (typically the λ, where absorbance is maximum represented by λ_{max}) is passed through a solution, then the quantity of light absorbed will be proportional to the concentration of the absorbing species in that solution. As the concentration of the solution increases, the amount of light absorbed will increase (Saunders et al., 1999).

Using the absorbance of a DNA sample at 260 nm and 280 nm, the ratio of the absorbances can be determined to assay the DNA quality and quantity for the previously extracted samples. A quartz cuvette will be used for these measurements, as it does not absorb in this region. The linear range is found between 0.1 and 1 absorption units, or about 5 to 50 µg/mL DNA. DNA preparations should be briefly vortexed or mixed by inversion and diluted to an appropriate concentration using 10 mM Tris-HCl or deionized water as a diluent. The instrument should be blanked with the same diluent. Using the absorptivity (or extinction) coefficient (a) for either double-stranded DNA, 50 µg/mL (A260 of 1 = 50 µg/mL DNA), or single-stranded DNA, 33 µg/mL (A260 of 1 = 33 µg/mL DNA) (Kallansrud and Ward, 1996), the absorption at 260 nm can be used to compute the concentration (c) of the sample using the Beer-Lambert Law, A_{260} = abc. The path length, b, is equal to 1 cm for a 1-cm cuvette. It is expected that the extracted DNA concentration from the simulated evidence samples will have a concentration at the low end of this range (0 to 500 ng/mL or 5 µg/mL) and that the concentration of buccal cells will be higher. If the sample is too concentrated and a dilution is necessary (e.g., absorbance exceeds 2.0) the concentration can be computed by multiplying the absorbance at 260 nm times the extinction coefficient and the dilution factor. For example, diluting 100 uL in 900 uL results in a 10-fold dilution and a dilution factor of 10. The overall DNA yield from the extraction can be computed by multiplying by the sample elution volume (e.g., 20 µL or 0.020 mL).

DNA concentration for double-stranded DNA (µg/ml) = (A260 nm − A320 nm) × 50 ug/mL × dilution factor

DNA concentration for single-stranded DNA (µg/ml) = (A260 nm − A320 nm) × 33 ug/mL × dilution factor

DNA yield (µg) = DNA concentration × eluted sample volume (ml)

The DNA purity or quality can be computed from the ratio of the absorbance at 260 nm and 280 nm (background absorbance may be subtracted using the absorbance at 320 nm) (Manchester, 1995).

$$Q = (A260 \text{ nm} − A320 \text{nm})/(A280 \text{ nm} − A320 \text{ nm})$$

If the 260/280 absorbance quotient (Q) is 1.7 to 2, the DNA is considered to be pure (Q = 2 for pure RNA). A lower absorbance ratio indicates impurity or degradation of the DNA. Phenol also absorbs strongly at 270 nm, and this impurity left over from the extraction method may falsely increase the calculated concentration. The Microcon/Centricon/Vivacon filtration step serves not only to concentrate the DNA but also to remove residual organics, including phenol, and other contaminants. Adding additional washes with 1X TE buffer prior to the elution step will reduce the phenol residue. Sample data are shown in Figure 6.1.

In this experiment, the quantity and quality of the extracted DNA will be determined using UV spectroscopy.

PROCEDURE

1. The spectrophotometer (e.g., the Agilent 8453 UV-Vis spectrophotometer) should be turned on to warm up for at least 10 minutes prior to any use. If the spectrophotometer has not been powered up when you enter the laboratory, turn it on. Do this by pushing the

FIGURE 6.1 Sample absorbance and DNA quality data.

DNA sample	A260 nm	A280 nm	Q (260 nm/280 nm)
1	0.04697	0.02683	1.75
2	0.05124	0.02764	1.85

power button on the spectrophotometer and then powering up the attached computer. Once the green light on the instrument appears, open the UV WinLabs software. Set up the software to read the UV absorbance at 260 nm, 280 nm, and 320 nm, and plot an absorption spectrum between 240 to 300 nm.

2. The spectrophotometer must be calibrated with a blank (e.g., water or 1X TE [10 mM Tris-HCl-0.1 mM EDTA] or 0.1X TE solution to which the DNA was eluted) prior to the analysis of your DNA solutions. Place blank solution in the quartz cuvette, and insert the cuvette into the cell holder in the instrument and collect the background. Do not place the liquid into the instrument without a cuvette. Select the button for "blank" to scan the blank.

3. Remove the blank solution from the cuvette, and replace it with one of the extracted DNA samples (or diluted to a 1:10, 1:100, or 1:1000 dilution, as necessary). Select the button for "sample" to scan the sample.

4. Record and tabulate the absorbance values at 260 nm, 280 nm, and 320 nm. Repeat step 3 until absorbances have been recorded for all samples.

5. Export sample spectrum from 240 to 300 nm for one of your samples as a .csv file for plotting in a graphing program. Note the excitation maxima.

6. Discard or store samples as desired or indicated by your instructor.

QUESTIONS

1. Tabulate your absorbance values, and calculate the DNA concentrations for your samples (taking into account the dilution) and quality values. Explain any deviations from the expected values.

2. What was the excitation maxima value for the DNA samples tested? Is this what you expected? Why or why not?

3. Which extraction method gave the highest yield?

References

Gallagher, S., 2011. Quantitation of DNA and RNA with Absorption and Fluorescence Spectroscopy. Curr. Protoc. Mol. Biol. 93, A.3D.1—A.3D.14.

Holden, M.J., Haynes, R.J., Rabb, S.A., Satija, N., Yang, K., Blasic Jr., J.R., 2009. Factors Affecting Quantification of Total DNA by UV Spectroscopy and PicoGreen Fluorescence. J. Agric. Food Chem. 57, 7221—7226.

Kallansrud, G., Ward, B., 1996. A comparison of measured and calculated single-and double-stranded oligodeoxynucleotide extinction coefficients. Anal. Biochem. 236, 134e—138e.

Labarca, C., Paigen, K., 1980. A simple, rapid and sensitive DNA assay procedure. Anal. Biochem. 102, 344e—352e.

Manchester, K.L., 1995. Value of A260/A280 ratios for measurement of purity of nucleic acids. BioTechniques 19, 208e—210e.

Saunders, G.C., Parkes, Helen C., 1999. Chapter 4: Quantification of Total DNA by Spectroscopy, Analytical molecular biology: quality and validation. Royal Society of Chemistry, 190.

Determination of DNA Quantity by Fluorescence Spectroscopy

OBJECTIVE

To learn how to use the fluorescence or luminescence spectrophotometer to determine the concentration of a DNA sample by recording emission intensity. A standard curve using serial dilutions of DNA of known concentration will be used to calibrate the emission intensity to derive the concentration of an unknown sample.

SAFETY

Handle the SYBR Green I dye carefully; SYBR Green I is an intercalating agent but is not noted to be mutagenic. Wear gloves; if you spill any chemicals on your person, wash them off immediately with soap and water.

Add all solutions to the cuvette, not directly to the luminescence spectrophotometer. Be careful not to spill solutions in the spectrophotometer. Clean up all spills immediately.

MATERIALS

1. Nuclease-free water or deionized water
2. Extracted DNA (5 to 50 ng range)
3. Micropipettes (0.5 to 10 μL, 10 to 20 μL, 10 to 200 μL, and 100 to 1000 μL) and tips
4. Perkin Elmer LS-50B luminescence spectrophotometer and quartz cuvette or BioRad iQ5 real-time PCR and 96-well plate or similar instrument
5. Lambda DNA (0.5 mg/mL stock, 20 mL diluted to 100 ng/mL) or calf thymus DNA
6. SYBR Green I (10,000X concentrate in DMSO)
7. Test tubes
8. Gloves, lab coats, and goggles

BACKGROUND

Light consists of packets of energy or radiation called photons. The energy of the photons varies with the wavelength of the light; light of a long wavelength contains less energy than light of a shorter wavelength. Photons in ultraviolet light, shorter wavelengths, have a higher energy than photons contained in visible light or longer wavelengths. Additionally, within the visible spectrum, red light has a longer wavelength and thus a lower energy than blue light. Compounds that absorb ultraviolet and visible light experience the displacement of certain electrons within the molecule from their ground energy state to an excited energy state. The ability of molecules to absorb light is dependent on their structure, and the absorption properties are reproducible. Compounds transmit irradiated light that is not absorbed. For example, if a solution is irradiated with white light and the substance absorbs red and yellow light, blue light is transmitted through the sample and can be observed by the eye.

In addition to absorbing UV light, some molecules such as pi-conjugated electron systems in dyes also fluoresce, or emit light that was previous absorbed. The emitted light is of a lower energy and of a longer wavelength in the visible region. The electrons undergo a transition from a higher energy excited state to a lower energy ground state when the electron relaxes. Non-radiative transitions may produce energy that is lost as heat and not converted to light energy.

In the luminescence spectrophotometer, the light from an excitation source passes through a filter or monochromator so that only selected wavelengths of light will irradiate the sample. The sample absorbs a proportion of this light, causing some of the molecules in the sample fluoresce. Further, the selected wavelengths of fluorescent light pass through a second filter, or monochromator, and reach a detector where fluorescence emission is recorded in relative fluorescence units (RFU). As the concentration of the solution increases, the amount of light emitted will increase (Gallagher 2011).

The advantages of fluorescence spectroscopy include that it is a rapid technique, instruments are widely available and no consumables or required reagents are needed. Disadvantages include that the fluorescent dye may not specific to human DNA depending upon label method used and that many substances fluoresce including metal ions, contaminating phenol from extraction procedures and some other organics.

An emission spectrum is obtained by plotting RFU as a function of wavelength (λ) (Figure 7.1) after exciting at a wavelength or wavelengths. A fluorescence emission spectrum is a chemical fingerprint and can be qualitatively used to identify molecular and atomic species present in a sample. It is more sensitive than absorbance alone in that all molecules do not fluoresce. It can be used to detect nanomolar concentrations of molecules. The difference between the excitation and emission maxima in the absorbance and emission spectra is referred to as the Stokes shift. SYBR Green I has a very narrow Stokes shift ($\lambda_{ex} = 497$ nm, $\lambda_{em} = 522$ nm).

A plot of RFU versus DNA concentration of lambda DNA is reported in Figure 7.2. The line equation can be used to calculate the quantity of an unknown using the fluorescence emission intensity of the unknown.

Each fluorescent molecule has a quantum yield, or the emission efficiency of a given fluorophore. The quantum yield is defined as the ratio (Φ) of the number of photons emitted to the number of photons absorbed:

$$\text{Quantum yield} = \Phi = \text{\# photon emitted}/\text{\# photons absorbed}$$

FIGURE 7.1 Fluorescence emission spectrum for SYBR Green I complexed with Lambda DNA ($\lambda_{EX} = 497$ nm, $\lambda_{EM} = 500\text{-}650$ nm) in FLWINLAB. DNA concentrations (top to bottom spectra): 38.3 ng/mL, 20 ng/mL, 10 ng/mL, 5 ng/mL, 2.5 ng/mL, 0 ng/mL.

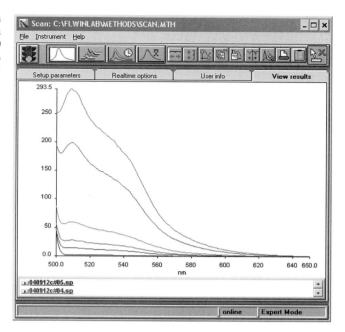

The maximum fluorescence quantum yield is 1. A quantum yield of 1 means that every photon absorbed results in a photon emitted. The quantum yield of SYBR Green I is 0.8. Proteins and DNA both fluoresce as their chemical molecules contain pi-conjugated electron systems. However, their quantum yields are relatively low compared to dye molecules used to detect DNA, including ethidium bromide, SYBR green I, PicoGreen, and Hoechst. When these intercalating agents embed themselves between stacked DNA bases, they exhibit a fluorescence enhancement following excitation with light. The dye molecule that intercalates with double-stranded DNA is excited and the emitted light is detected in the experiment. The fluorescence emission of an intercalating agent in the presence of DNA is enhanced

FIGURE 7.2 Graph of relative fluorescence units (RFU) versus concentration for lambda DNA in Figure 7.1.

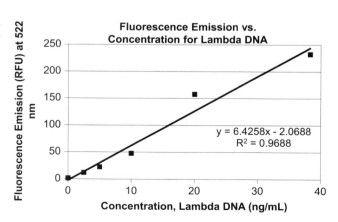

TABLE 7.1   BioRad iQ5 Excitation and Emission Filter Positions for fluorescent dyes

Position	Excitation (nm)	Emission (nm)	Dye
2	485	530	Fluorescein (FAM), SYBR Green I
3	530	575	HEX, TET, VIC, JOE
4	545	585	TAMRA/Cy3
5	575	625	Texas Red, ROX
6	630	685	Cy5, LC640

from the fluorescence emission observed from the unbound molecule (Holden et al., 2009, Rengarajan et al., 2002, Singer et al., 1997, biotek.com).

Hoechst (bis benzimide) dyes are sensitive fluorescent nucleic acid stains that are somewhat selective for double-stranded DNA, do not show significant fluorescence enhancement in the presence of proteins, and allow for the detection and quantitation of DNA to 10 ng/mL concentration. PicoGreen (25 pg/mL to 1 ng/mL) and SYBR Green I dyes are reported to be 1000 times more sensitive than ethidium bromide (100 ng/mL to 10 ug/mL) or Hoechst 33258 and can accurately measure DNA concentrations down to 0.5 ng/mL. However, SYBR Green I is substantially less expensive than PicoGreen (Gallagher, 2011, Holden et al., 2009, Rengarajan et al., 2002, Vizthum et al., 1999, Singer et al., 1997).

In this experiment, the extracted DNA will be quantified using SYBR Green I dye ($\lambda_{ex} = 497$ nm, $\lambda_{em} = 522$ nm) and a solution fluorescence spectrometer or a real-time PCR or plate reader instrument. Table 7.1 shows the excitation filter position for various dyes in the BioRad iQ instrument. Position 2 would be the best for SYBR Green I. Phenol also absorbs in the UV region at 270 nm but exhibits a fluorescence emission at approximately 290 nm, well outside of the fluorescence emission to be recorded for SYBR Green I at 522 nm.

In this experiment, we will determine the concentration of the extracted DNA using lambda DNA or calf thymus DNA standards at various dilutions of known concentrations to calibrate the emission intensity using SYBR Green I dye as a fluorescence reporter. The concentration of the unknown will be computed by the line equation from a graph of RFU versus the concentration of DNA standard as shown in Figure 7.2. The simulated evidence samples are on the low end of the 0–500 ng/mL range but the extracted DNA from buccal cells will be higher and will need to be diluted. If the RFU values exceed 1000, the samples need diluting. Multiply by the dilution factor when determining the concentration of the extract.

PROCEDURE

Part A: Preparation of Solutions

1. In a microcentrifuge tube or small test tube, obtain 5 mL of the diluted lambda DNA or calf thymus DNA standard (diluted to 100 ng/mL from approximately 0.5 mg/mL stock) provided.

2. Obtain six more clean, sterile test tubes or PCR tubes. Dilute DNA serially to make standards to the following concentrations: 0 ng/mL, 3.625 ng/mL, 7.5 ng/mL, 12.5 ng/mL, 25 ng/mL, and 50 ng/mL using $M_1V_1 = M_2V_2$. (Make a table of the dilutions and how much to pipette.) Dilute solutions using sterile, autoclaved, deionized water to create a final volume of 4 mL (or the volume of the quartz cuvette) of the appropriate concentration for option 1 or a final volume of 50 μL for option 2, as directed by your instructor.

3. Using micropipettes, pipette the appropriate amount of DNA into the corresponding tube.

4. Add 0.64 uL for option 1 (0.008 μL for option 2, dilute to 100X to pipet 0.8 μL) of the SYBR Green I dye for a final dilution of 1:6250 (6250-fold) to each tube. Cap each tube, mix well, and incubate in the dark for 5 minutes.

5. For the blank (0 μg/mL), obtain an unknown in a microcentrifuge tube and add deionized water to the cuvette volume. Cover the tube and mix well.

6. Prepare unknowns using the same method as in steps 3 to 4 for standard samples.

Part B: Detection of Double-stranded DNA

Option 1: Luminescence Spectrophotometer (Perkin Elmer LS 50B or similar instrument)

1. The spectrophotometer needs to warm up for at least 30 minutes prior to any use. If the spectrophotometer has not been turned on when you enter the laboratory, turn it on. Turn on the computer and open FL Win Lab.

2. Select scan. Set the excitation wavelength to 450–500 nm and emission wavelength to 500 to 650 nm to record excitation and emission spectra (5 nm excitation and emission slit widths, 240 nm/min) for a 3D excitation-emission plot. For the concentration determination using the DNA standards, set the excitation to 497 nm and emission to 522 nm.

3. Obtain a 4-mL quartz cuvette. Place most dilute standard-SYBR Green I solution in the cuvette and scan. Record the relative emission intensity. Repeat, increasing to the most concentrated standard. Rinse the cuvette 10 times with deionized water between samples.

4. Scan your "unknown" extracted DNA solutions. Record relative emission intensity for each sample.

Option 2: Plate Reader or Real-Time PCR Instrument (BioRad iQ5 or similar instrument)

1. Mix samples by pipetting up and down.

2. Pipette your samples into a 96-well plate; carefully record wells used. Pipette the sample onto the side of the well, and do not release the pipette until withdrawn from the sample so as to not produce bubbles. Load wells A to G for standards and an "unknown" extracted DNA sample in H.

3. When all standard samples and unknowns have been loaded, use the plastic sealing squeegee to seal Microseal plastic covers over all samples.

4. Open the PCR machine for the instrument and insert the well plate; fully close the lid.
5. Turn on the computer attached to the BioRad iQ5 RT-PCR instrument. Turn on the BioRad iQ5 RT-PCR instrument; wait for the self-test to complete (approximately 10 minutes prior to use).
6. When the User Log-on screen appears, open the iQ5 v.2.0 software on the computer.
7. Under the Setup tab, select Protocol and select Create New to set up a new experiment. In the table at the bottom of the screen, edit the cycle settings to read: Cycle 1, Step 1, 0:15 minute Dwell time, 25.0 Setpoint. Select Save and Exit Protocol Editing, give it a filename, and click save.
8. From the Setup page, now select Plate. Create New to input student sample information. Click on the positive control (lambda or calf thymus DNA sample) symbol, and click in the column and row desired (e.g., A1-G1) to insert positive control samples. Click on the box with the X in it to identify the extracted DNA "unknown" samples, and click in the column and row desired (e.g., H1) to label the samples. Clicking on the boxes highlights the sample for editing in the table below, and student sample identifiers can be added. When completed, select Save and Exit Plate Editing and give the plate a filename.
9. In the top right-hand corner of the screen, select Run. The instrument is calibrated with the five dyes; select Use Persistent Well Factors for dye deconvolution. Select Begin Run and give it a data filename. The selected protocol and plate will have the filenames displayed to ensure proper run parameters.
10. When the run is completed, click OK in the box on the screen. The run saves automatically to the preset filename you gave.
11. To analyze the data, open the Data File. The first screen is the PCR Quant with all samples plotted for RFU versus cycle number. Clicking on threshold cycle yields the entrance to log phase for the setting given (move the green line to adjust). By clicking Display Wells, wells may be turned off or on for viewing by selecting the well to turn off and clicking OK.
12. Record the RFU values for all standards and unknowns.

Part C: Excitation-Emission Spectrum Plot (if performing Option 1)

1. Measure and record the excitation and emission spectra of your extracted DNA by scanning over the wavelength range from 450 to 500 nm and the emission from 500 to 650 nm.
2. Print the spectra or save the data as a .csv file. Record the observed excitation and emission maxima. Sketch graphs in your lab notebook.

QUESTIONS

Tabulate the emission data (in RFUs) at the concentrations of lambda DNA (or other standard) you used. Calculate the concentration of the extracted DNA using a plot of relative fluorescence emission intensity of SYBR Green I versus concentration (ng/mL) for the DNA standards using the slope of the regression line. Report your excitation and emission spectrum plot for an extracted (unknown) DNA sample, and label the excitation and emission maxima.

References

Gallagher, S., 2011. Quantitation of DNA and RNA with Absorption and Fluorescence Spectroscopy. Curr. Protoc. Mol. Biol. 93, A.3D.1–A.3D.14.

Holden, M.J., Haynes, R.J., Rabb, S.A., Satija, N., Yang, K., Blasic Jr., J.R., 2009. Factors Affecting Quantification of Total DNA by UV Spectroscopy and PicoGreen Fluorescence. J. Agric. Food Chem. 57, 7221e–7226e.

Rengarajan, K., Cristol, S.M., Mehta, M., Nickerson, J.M., 2002. Quantifying DNA concentrations using fluorometry: A comparison of fluorohores. Molecular Vision 8, 416–421.

Singer, V.L., Jones, L.J., Yue, S.T., Haugland, R.P., 1997. Characterization of PicoGreen reagent and development of a fluorescence-based solution assay. Anal. Biochem. 249, 228–238.

Vitzthum, F., Geiger, G., Bisswanger, H., Brunner, H., Bernhagen, J., 1999. A Quantitative Fluorescence-Based Microplate Assay for the Determination of Double-Stranded DNA Using SYBR Green I and a Standard Ultraviolet Transilluminator Gel Imaging System. Analytical Biochemistry 276, 59–64.

www.biotek.com/resources/articles/dna-hoechst-33258.html, accessed 11-22-2011.

8

Real-Time Polymerase Chain Reaction (PCR) Quantitation of DNA

OBJECTIVE

To learn how to set up and run real-time PCR reactions to determine the concentration of an extracted human DNA sample.

SAFETY

Wear gloves. Handle the SYBR Green I reaction mix carefully. SYBR Green I is an intercalating agent. Immediately wash off any chemicals with which you come into skin contact using soap and water.

The real time PCR instrument is very sensitive to background noise. Do not touch the plate without gloved hands. To avoid contamination, always wear gloves when handling the samples and reagents.

MATERIALS

1. BioRad iQ 96-Well PCR Plates or equivalent
2. BioRad Microseal "B" film or equivalent
3. BioRad 2x iQ SYBR Green SuperMix or equivalent
4. Nuclease-free water
5. Extracted DNA template (1 to 10 ng range)
6. TPOXF and TPOXR primers (diluted to 5 μM) (IDT, from 25 nmole, standard desalting)
   Forward: 5'-CGGGAAGGGAACAGGAGTAAG-3'
   Reverse: 5'-CCAATCCCAGGTCTTCTGAACA-3'
7. 10 ng/μL K562 DNA diluted to 0.100 ng, 0.500 ng, 1 ng, 2 ng, 5 ng and 10 ng and 20 ng samples

8. BioRad iQ5 RT-PCR instrument with BioRad iQ5 software v. 2.0 or equivalent
9. Micropipettes (e.g., 0.5 to 10 μL, 1 to 20 μL, 10 to 200 μL, and 100 to 1000 μL) and autoclaved tips
10. Sterile 1.5 mL microcentrifuge tubes
11. Gloves
12. Freezer
13. Microwave or heating plate
14. Agarose
15. Gel box and power supply
16. 100X SYBR Green I
17. 6X gel loading dye
18. DNA ladder
19. Photodocumentation system with SYBR Green I filter

BACKGROUND

The polymerase chain reaction (PCR) was invented by Kary Mullis in 1983. PCR can be used to copy DNA extracted from samples. This allows forensic scientists to type DNA using minute, diluted, and degraded DNA samples. As PCR is highly specific, contaminating DNA from fungal, bacterial, plant, or even other animal sources will not be amplified and only the human DNA template will be amplified.

The real-time polymerase chain reaction (RT-PCR), also known as quantitative real-time polymerase chain reaction (Q-PCR/qPCR/qrt-PCR) or kinetic polymerase chain reaction (KPCR), amplifies and simultaneously quantifies a specific region of a targeted DNA molecule. RT-PCR detects DNA as the chain reaction occurs—that is, in real time. (Higuchi et al., 1993). Traditional PCR requires the user to wait until the experiment is completed and the samples are separated by gel or capillary electrophoresis to view the results. This real-time expression allows the forensic scientist to analyze the data using computer software that calculates the relative gene expression or quantity of DNA amplicons produced in several samples at once.

To run a PCR reaction, a DNA template, DNA polymerase, 5' and 3' primers complementary to regions upstream and downstream of the DNA locus of interest (Figure 8.1),

TPOX NCBI Nucleotide Sequence
(Accession NG_011581)

```
           ⇨  5' PRIMER
  81261   GCGGGAAGGG AACAGGAGTA AGACCAGCGC ACAGCCCGAC TTGTGTTCAG AAGACCTGGG
          CGCCCTTCCC TTGTCCTCAT TCTGGTCGCG TGTCGGGCTG AACACAAGTC TTCTGGACCC
                                                                3' PRIMER
  81321   ATTGG
          TAACC
           ⇦
```

FIGURE 8.1   PCR primer locations (bold) to be used to amplify *Homo sapiens* thyroid peroxidase (TPO) on chromosome 2, NCBI Nucleotide Sequence (Accession: NG_011581) (64 bp) (Horsman et al., 2006). The sequence is shown from 5' to 3' on the top strand.

magnesium ions, bovine serum albumin, and Tris buffer are needed. A thermocycler is used to perform temperature cycling. The heat-stable DNA polymerase was originally isolated from *Thermus aquaticus* thermophilic bacteria and can be used in temperature-cycling reactions without denaturing the enzyme. The optimal temperature for the *Taq* polymerase is 72° C, but it is active over a large temperature range and will begin working as soon as the PCR primers anneal, or bind, to the complementary portion(s) of the DNA. As a result, real-time PCR typically employs only a two-temperature cycle in the reaction: an annealing/extension step and a denaturation step. Keeping the mixture cold prevents the enzyme from extending the primers and forming artifactual primer-dimers, hydrogen bonding interactions of the primers with themselves, prior to initiating the reaction. Primers that are hydrogen-bonded to other primers are not available for binding to the DNA template. PCR is sensitive to small changes in buffers, ionic strength, primer concentrations, and choice of thermal cycler and thermal cycling conditions. Typical concentrations are shown in Table 8.1. All of these can affect the success of PCR amplification and must be optimized for optimal results with untested primers. Further, contamination and PCR inhibitors such as indigo from blue jeans, heme from hemoglobin, humic acids from soils, and EDTA can also affect the DNA amplification. Extreme care must be taken to avoid cross-contamination in preparing DNA samples, when handling primer pairs, and setting up amplification reactions when working with human DNA. Reagents and materials used prior to amplification should be stored separately from those used following amplification. A negative (no template) control reaction in which nuclease-free water is substituted for extracted or purchased DNA should always be performed to ensure reagent purity.

Standard PCR reaction volumes are 20 to 50 µl but 5 to 100 µL may also be used. For a larger reaction volume, it takes a longer time interval to achieve the desired heating and cooling temperatures. Low volumes pose difficulty with pipetting such small volumes. PCR may be conducted in thin-walled PCR tubes or 96-well plates. The number of cycles performed

TABLE 8.1 Typical PCR Reagent Concentrations

Reagent	Optimal Concentration	Function
Tris-HCl, pH 8.3 (25 °C)	10-50 mM	pH Buffer for optimal reaction conditions
$MgCl_2$	1.5-2.5 mM	Stabilize DNA and dNTPs
KCl	50 mM	Stabilize reaction
dNTPs	200 µM each dNTP	Building blocks to add to primers to make new DNA amplicon products
Taq DNA polymerase	0.5-5 U	Catalyzes the formation of phosphodiester bonds between the growing primer chain and the dNTPs, cleaves phosphoanhydride bond of dNTPs
Bovine Serum Albumin (BSA)	100 µg/mL	Stabilizes reaction, binds PCR inhibitors
Primers (each, 5' & 3') (18-35 base pairs length)	0.1-1.0 mM	Single-stranded DNA elements that bind denatured DNA template so that DNA polymerase can bind and extend the primer
Template extracted DNA	1-2.5 ng	Source of sequence to extend from primers

in PCR may vary, but 28 cycles is standard for 1 ng of template DNA, whereas up to 42 cycles may be used for extremely low concentration DNA.

In real-time PCR, the amplified DNA is detected using fluorescence reporters that intercalate, or slide between, stacked DNA base pairs (without preference for sequence) or via fluorescent reporters attached to the 5′ end of the 5′ (top) primer or ddNTP in mini-sequencing. In this experiment, the intercalating agent SYBR Green I dye (λ_{ex} = 497 nm, λ_{em} = 522 nm) will be used to detect the DNA. In this laboratory, a single set of primers (one TPOX 5′ and one 3′ primer; Figure 8.1) published for this purpose will be used to amplify the extracted DNA template and the purchased DNA template of known quantity. The optimal concentration for PCR is 1 to 2.5 ng of template DNA. Without a prior estimation of quantity, it is reasonable to use 1 μL of a 1:10 or 1:100 dilution of the Chelex or PCIA-extracted genomic DNA templates. Alternatively, the DNA template may be diluted to this concentration based on the previously calculated concentrations using the gel, fluorescence or UV methods in previous experiments. The size of the expected amplicon is 64 base pairs; its production will be determined by the melting temperature post-PCR (and may also be confirmed using a 1% agarose gel loaded with a DNA size ladder). A typical amplification plot for RFU versus cycle is shown in Figure 8.2.

Real-time PCR also permits a calculation of the cycle number for which a sample enters the log or exponential growth phase for a reaction. By setting a relative fluorescence unit (RFU) value for comparison for all samples, the threshold cycle to cross the RFU can be identified for all samples. A plot of threshold cycle (C_T) versus log of the DNA concentration of K562 DNA standards (Figure 8.3) can be used to determine the concentration of an unknown DNA template by inputting the C_T into the line equation. A 3.33 cycle increase in amplification time indicates a 10-fold dilution in DNA concentration. Good-quality results have a R^2 of 0.98 or better and an amplification efficiency (E) of 90% to 110%.

$$E = 10^{(-1/\text{slope})-1}$$

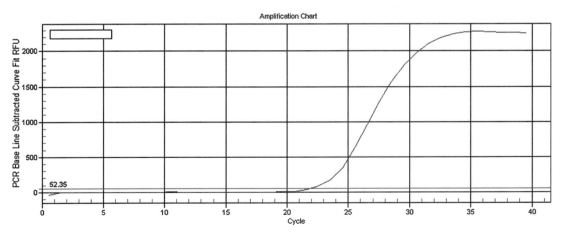

FIGURE 8.2   Exponential amplification curves that show sample results of 40 cycles of real-time PCR with 1 ng K562 DNA template using TPOX primers. Curves of K562 DNA as detected by SYBR Green I.

FIGURE 8.3 Sample K562 human DNA standard curve: C_T versus log DNA concentration.

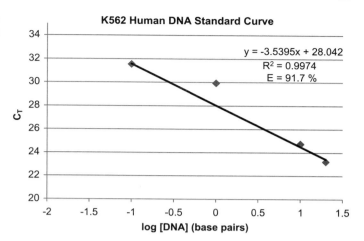

The size of the amplicon can also be estimated. The RT-PCR permits a DNA melt analysis. The amplified DNA is subjected to 0.5° C increases in temperature from 50° C to 95° C. An inflection point in a plot of RFU versus temperature (or peak in a first derivative of the RFU versus temperature) can be used to determine the melting temperature of the amplicon (Figure 8.4).

The website, www.basic.northwestern.edu/biotools/oligocalc.html#helpbasic, can be used to predict the melting temperature of DNA amplicons. The sequence of the 64 bp amplicon (see Figure 8.1) was calculated to be 79° C (basic), 93° C (salt adjusted with 50 mM from mix), and 82° C (nearest neighbor). The basic melting temperature calculations are provided as a baseline for comparison; they are the least reliable but most often used. The calculation can be used to assess if the amplicon is produced. The melting temperature can be affected by

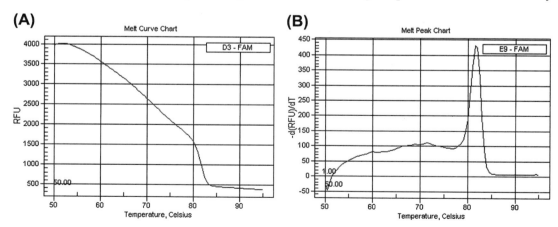

FIGURE 8.4 A. Sample melt curve plot from 50° to 95° C showing the temperature inflection at 81° C for the K562 DNA reaction with the TPOX primers shown in Figure 8.2 (A). B. First derivative plot of the melting curve from 50° to 95° C showing the predominant peak at 81° C for the K562 DNA (B).

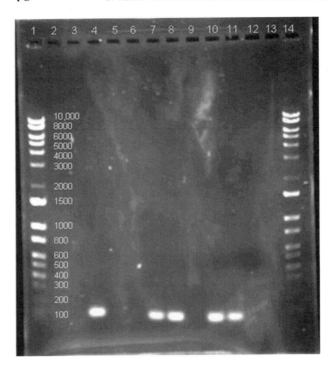

FIGURE 8.5 Sample 1% agarose gel of PCR of K562 DNA using TPOX primers. Lane 1 contains the DNA Logic ladder (sizes indicated), Lanes 3, 7, 8, 10, and 11 contain the 64 bp amplicon.

the presence of detergents, other counter ions, solvents (e.g., ethanol), and denaturants (e.g., formamide).

In a second week, an agarose gel can also be used to confirm the production of the expected PCR product (Figure 8.5). A 1% agarose gel can be used to compare the migration of the amplicon with that of the DNA size fragments in a DNA ladder. Careful measurements of the ladder can be used to produce a graph of log base pairs versus the distance traveled from the well that can be used to calculate the size of the amplicon.

In this experiment, the TPOX locus primers will be employed to amplify various concentrations of K562 (or other) DNA standard to produce a standard curve and DNA extracted in a previous experiment will be quantitated.

PROCEDURE

1. Pipet the following into separate 1.5 mL microcentrifuge tubes for the extracted DNA samples:
   12.5 μL of 2x iQ SYBR Green SuperMix
   1 μL of 5 μM TPOXF
   1 μL of 5 μM TPOXR
   9.5 μL of nuclease-free water
   1 μL of extracted template DNA (1 to 2.5 ng) or K562 DNA (0.1 ng to 20 ng per μL)
   25 μL Total Reaction Volume

Prepare at least one negative control sample (substitute nuclease free water for template DNA so add 10.5 μL of nuclease-free water) for the plate. Prepare one set of positive control samples (substitute K562 DNA, range diluted to 0.1 ng, 0.5 ng, 1 ng, 2 ng, 5 ng, 10 ng, and 20 ng/μL for template DNA) (individually or as a class).

2. Mix samples by pipetting up and down gently so as not to introduce bubbles. Centrifuge samples briefly.

3. Pipette your samples into the 96-well plate by pipetting down the sides of the plate (or PCR tube); carefully record wells used. Pipette sample to the side of the well, and do not release pipette until withdrawn from the sample so as to not produce bubbles.

4. When all samples have been loaded to the 96-well plate, use the plastic sealing device to seal Microseal plastic covers over all samples.

5. Open the PCR machine for the instrument and insert the well plate; fully close the lid.

6. Turn on the computer attached to the BioRad iQ5 RT-PCR instrument. Turn on the BioRad iQ5 RT-PCR instrument; wait for the self-test to complete (approximately 10 minutes prior to use).

7. When the User Log-on screen appears, open the iQ5 v.2.0 software on the computer.

8. Under the Setup tab, select Protocol and select Create New to set up a new experiment. In the table at the bottom of the screen, edit the cycle settings to read: Cycle 1, Step 1, 3:00 minute Dwell time, 95.0 Setpoint. After Cycle 1 is created, click on the ... under Options on the Cycle 1 line to insert cycles "After" and select "2 Steps" and select insert. Edit Cycle 2, Step 1 to read: 0.15 Dwell time at 95.0 Setpoint and Cycle 2: Step 2 to read: 1:00 Dwell time and 60.4 Setpoint. After Cycle 2 is created, click on the ... under Options on the Cycle 2 line to insert cycles "After" and select "1 Step" and select insert. In the drop-down box under PCR/Melt Data Acquisition, select Melt Curve. Edit the preset conditions to read: 50.0 Setpoint. This will execute a melting curve of 91 cycles for 30 seconds each of melting every 0.5° C from 50° C. Select Save and Exit Protocol Editing, give a filename, and click save (Figure 8.6).

9. From the Setup page, now select Plate. Create New to input student sample information. To insert a negative control, click on the negative control icon in the column and click on the box in the row desired (e.g., A1). Click on the positive control symbol and click in the column and row desired (e.g., B1 to H1) to insert a positive control sample. Click on the box with the X in it to identify the student samples and click in the column and row desired (e.g., A2 through H2) to denote the student samples. Clicking on the boxes highlights the sample for editing in the table below and student sample identifiers can be added. When completed, select Save and Exit Plate Editing and give the plate a filename (see Figure 8.6). The completed plate and run files are shown in Figure 8.6.

10. In the top right-hand corner of the screen, select Run. The instrument was previously calibrated with the 5 dyes, so we select Use Persistent Well Factors for dye deconvolution. Select Begin Run and give a data filename. The selected protocol and plate will have the filenames displayed to ensure proper run parameters.

11. When the run is completed, click OK in the box on the screen. The run saves automatically to the preset file.

12. To analyze the data, open the Data File. The first screen is the PCR Quant with all samples plotted for RFU versus cycle number. Clicking on threshold cycle yields the entrance

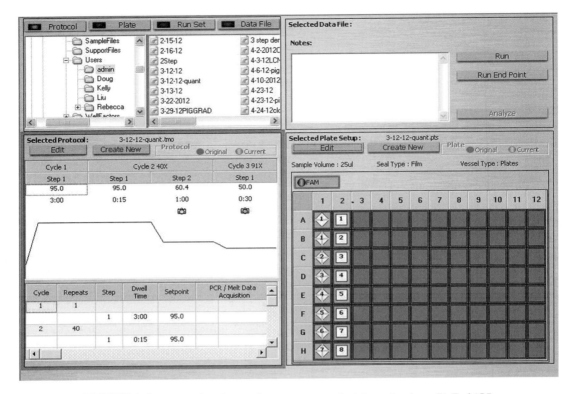

FIGURE 8.6   Setup of cycling and temperature and plate protocols on BioRad iQ5.

to log phase for the setting given (move the green line to adjust). By clicking Display Wells, wells may be turned off or on for viewing by selecting the well to turn off and clicking OK.

13. The melt data can be analyzed in the Melt Curve Peak window. The left-hand plot is the RFU versus temperature (°C). The right-hand plot is the first derivative of the RFU versus temperature (°C). The RFU values and melt temperatures for the peaks are given in the table at the bottom of the screen.

14. The graphs can be exported using PDF Converter. Alternatively, the graphs can be copied to Paint for saving by right-clicking in the graph and selecting copy graph and copying to Paint. Files can then be saved as .JPG or .TIFF files.

15. Optional second week. Prepare a 1% agarose gel (Refer to Experiment 5) using 0.5 g of agarose melted in 50 mL 1X TAE buffer and set in a gel box with 14-well comb. Add 3 μL of the 6x loading buffer per 25 μL sample. Combine 8 μL of the DNA Logic ladder with 1 μL of 100x SYBR Green I dye and 1 μL of the 6x loading dye. Load 12 μL samples and all of ladders while carefully RECORDING LOCATIONS of samples. Run at 150 V for a 10 cm gel for 30 minutes. Remove gel from gel box and record image with photodocumentation unit (see also Chapter 5).

QUESTIONS

1. What were the expected/calculated and observed T_m values for the amplicons? Give reasons for any deviations.
2. State the threshold cycle for your samples. What is the relevance of this value?
3. Based on comparison with the K562 positive control, estimate the concentration in your original DNA template samples by plotting the threshold cycle (C_T) versus log of the known DNA concentrations.

References

Higuchi, R., Fockler, C., Dollinger, G., Watson, R., 1993. Kinetic PCR analysis: real-time monitoring of DNA amplification reactions. Biotechnology (NY) 11, 1026–1030.

Horsman, K.M., Hickey, J.A., Cotton, R.W., Landers, J.P., Maddox, L.O., 2006. Development of a human-specific real-time PCR assay for the simultaneous quantitation of total genomic and male DNA. J. Forensic Sci. 51, 758–765.

Multiplex Polymerase Chain Reaction (PCR) Primer Design (*in Silico*)

OBJECTIVE

To learn how to design polymerase chain reaction PCR primer multiplexes for short-tandem repeat (STR) deoxyribonucleic acid (DNA) sites of interest.

SAFETY

No special safety precautions. This is a computer-based *in silico* lab.

MATERIALS

1. Computer with an Internet connection

BACKGROUND

The polymerase chain reaction (PCR) is a method that can be used to copy DNA regions of interest using DNA polymerase. Forensic DNA biologists rely heavily on PCR-based methods for copying and analyzing DNA. However, many crime scenes produce multiple biological samples that require painstaking analysis. Multiplex kits, in which multiple pairs of PCR primers are added into the same PCR reaction mix, facilitate this simultaneous analysis of multiple samples. This innovation saves the crime labs tremendous time and money. Getting multiple reactions to proceed in 25 μL is not a trivial task. The concentrations of the PCR primers, 2′-deoxynucleotide triphosphates (dNTPs), DNA polymerase, magnesium, bovine serum albumin (BSA), KCl, and Tris reagents must all be optimized so that approximately equal quantities of the PCR products are produced in the multiplex reaction. This can take considerable time.

The unique reagent in the multiplex kits is the PCR primer. The primers are short DNA oligonucleotides that are complementary and specific to a target region, or locus, of the

DNA of interest. The primers are absolutely necessary, as the DNA polymerase needs a double-stranded DNA segment or handle to bind to in order to catalyze new phosphodiester bonds between bases in the primer and new bases that form complementary base pairs to the template. In forensics, these oligonucleotides probes usually bind upstream and downstream of an autosomal STR used for DNA typing (Table 9.1). However, the target may also be the X-chromosome, Y-chromosome, and mitochondrial DNA chromosome SNPs and STRs.

The U.S. Federal Bureau of Investigation (FBI) uses 13 STR sites that are routinely evaluated and inputted to its Combined DNA Index System (CODIS) database for convicted criminals, casework evidence, and even reference samples and unknown samples from missing persons cases (see Table 9.1). There are also many other non-CODIS STR sites that have been identified in the human genome, which can be used for identity typing and which have been adopted outside of the United States.

In the experiment, you will design PCR primers that will amplify one of the CODIS loci shown in Table 9.1 (Elkins, 2011, Lima and Garces, 2006, Kim, 2000). The upstream and downstream sequences will be obtained from the NCBI website using the NCBI accession number.

DNA amplification using the PCR reaction requires the use of two primers. These primers anneal specifically to their target sequences and serve as "grips" or "handles" for the DNA polymerase to latch onto. The two primers bind to the top and bottom strands, or 5' and 3' strands, respectively, of the antiparallel double-stranded DNA helix of the DNA region of

TABLE 9.1 NCBI Accession Numbers and STR Repeats of CODIS Loci

NCBI accession number	Locus	Allele	STR Repeat
X14720 or U63963	CSF1PO	12	AGAT
D00269	TH01	9	AATG
M68651	TPOX	11	AATG
AC008512 or G08446	D5S818	11, 11	AGAT
AC004848 or G08616	D7S820	13, 12	GATA
AF216671 or G08710	D8S1179 (formerly D6S502)	13, 12	TCTA
AL353628 or G09017	D13S317	11, 13	TATC
AC024591 or G07925	D16S539	11, 11	GATA
AP001534 or L18333	D18S51	18, 13	AGAA
M25858	vWA	18	TCTA[TCTG]3-4[TCTA]n
NT_005997	D3S1358	18	TCTA[TCTG]2-3[TCTA]n
M64982	FGA	21	[TTTC]3TTTT TTCT[CTTT]nCTCC [TTCC]2
AP000433	D21S11	29	[TCTA]n[TCTG]n{[TCTA]3TA[TCTA] 3TCA[TCTA]2TCCATA}[TCTA]nTA TCTA

interest, as assayed by Blast. The primers are termed the 5', or forward, and 3', or reverse, primers, respectively. The primers bind the DNA by complementary base pairing (A = T and C ≡ G). DNA polymerase then functions to extend the primer sequence into the STR or sequence of interest by adding the complementary deoxynucleotide triphosphates or dNTPs (e.g., dATP, dTTP, dCTP, and dGTP) to the oligonucleotide primer based on the sequence of the target strand.

For real-time PCR, the best primers are those that amplify a short sequence (200 bp or less), but primers that amplify less than 500 bp routinely give results. In addition, to sequence a PCR product, the primers should be ≤500 bases apart so that the product is no longer than approximately 500 base pairs, a typical readable sequence using a single set of sequencing primers. The primer must be complementary to the region upstream of the beginning of the target sequence and the reverse complement of the region downstream of the target.

The primers themselves are designed together to optimize the binding and amplification potential. The characteristics of the primers (Table 9.2) can be evaluated computationally using OligoAnalyzer and other software such as AutoDimer (Vallone and Butler, 2004). The primers must possess similar melting/annealing temperatures of 55° to 72° C to ensure specific priming (and less than or equal to a 5° C difference in melting temperatures, T_m) and have a 40% to 60% GC content to achieve the melting temperature. They should be 18 to 30 bases in length; shorter primers often have low melting temperatures and are not specific enough for the target. This is sufficiently long enough for specific binding to the complementary target yet also cost effective, as primer prices reflect the number of nucleotide bases. The shorter the primer, the higher the concentration of the oligonucleotide (oligo) in a given micromolar sample purchase. The primers must be structurally compatible and not interact with each other or themselves to form "primer dimers" or "hairpin" structures so there should be no more than three identical consecutive bases. As the guanine and cytosine (GC) base pair has three hydrogen bonds, the melting temperature of that pair will be higher than that of adenine and thymine (AT) with two base pairs. To achieve a higher melting

TABLE 9.2   Characteristics of Optimal PCR Primers (Butler, 2005)

Feature	Optimal Condition
Primer length	18-30 bases
Primer T_m	55-72 °C
GC content (%)	40-60 %
T_m difference for forward/reverse primer pair	≤ 5 °C
Distance between primers on target DNA	< 500 bases apart
No self-complementarity (hairpins)	≤ 3 continguous bases
No primer dimer, especially at 3' ends	≤ 3 continguous bases
No long runs with the same base	< 4 contiguous bases
Unique oligonucleotide sequence	Best/only match in Blast search

temperature, shift to GC-rich sequences. The increase in hydrogen bonds will also promote the specificity of the sequence (Butler, 2005, Mitsuhashi, 1996).

The United States relies on the use of multiplex PCR to simultaneously amplify the 13 Combined DNA Index System (CODIS) STR loci of interest in one test tube and expeditiously complete forensic casework. In the DNA typing kits sold by manufacturers such as Applied Biosystems (ABI), Promega, and other manufacturers, multiple primer sets are added to the reaction mixture with the DNA template. These primers have the characteristics of all being complementary to different target sequences and have little, if any, self-complementarity or complementarity to any of the oligos in the set, have separate binding sites on the target DNA, and have no long runs of contiguous bases. Promega has published its primers (Masibay et al., 2000). In multiplex PCR, the extension times are often increased so that the polymerase can fully copy all targets. For multiplex analysis of the STRs on the genetic analyzer, the primers have been adjusted (by moving them up or downstream of the repeat or adding a non-nucleotide linker) to achieve good size separation between loci products labeled with the same fluorescent dye and to not produce nonspecific products that might make the DNA profile difficult to read. Similar expected amplicon sizes are labeled with different fluorescent dyes (or can be separated using post-PCR gel electrophoresis).

In this experiment, you will design one set of PCR primers to amplify a CODIS STR locus and then test the primers with the primers designed for a second locus for heterodimer complementarity between the primers. You will also use a program, Blast (www.ncbi.nlm.nih.gov/blast), to test that the primers you design are specific for only one locus in the human genome (that they are specific).

PROCEDURE

Part A: Obtaining an STR DNA Sequence from the NCBI Website

1. Access www.ncbi.nlm.nih.gov.
2. Select Search: Nucleotide for the accession number corresponding to the CODIS STR of interest (see Table 9.1). Each person should select a different STR or DNA gene of interest to design a primer. Record the locus, accession number, and repeat for the selected STR locus.
3. To find the nucleotide sequence, scroll to the bottom of the page to the word ORIGIN, select the one-letter nucleotide code sequence that follows it, and paste the sequence to a text or Microsoft Word file. The copied sequence is the top (5' to 3' strand).

Part B: Preliminary Primer Design

1. Locate the STR repeat in the DNA sequence. Use the Find function in the web browser or Microsoft Word to enter the repeat and search for the expected number of repeats in tandem.
2. Working directly upstream from the repeat, locate a GC-rich segment without long stretches of a same base. Copy an 18- to 30-nucleotide segment of this region and paste it into the text file. This is a preliminary 5'-primer.

FIGURE 9.1 TPOX NCBI Nucleotide Sequence (*Accession* M68651): AATG STR repeat (underlined), and sample designed primers (arrows and bold).

```
                                                       5' PRIMER  ┌──────────────────────────►
1801 GTTTCAGGGC TGTGATCACT AGCACCCAGA ACCGTCGACT GGCACAGAAC AGGCACTTAG
1801 CAAAGTCCCG ACACTAGTGA TCGTGGGTCT TGGCAGGTGA CCGTGTCTTG TCCGTGAATC

     STR REPEATS  1    2     3     4     5     6     7     8     9    10    11
1861 GGAACCCTCA CTGAATGAAT GAATGAATGA ATGAATGAAT GAATGAATGA ATGAATGTTT
1861 CCTTGGGAGT GACTTACTTA CTTACTTACT TACTTACTTA CTTACTTACT TACTTACAAA

1921 GGGCAAATAA ACGCTGACAA GGACAGAAGG GCCTAGCGGG AAGGGAACAG GAGTAAGACC
1921 CCCGTTTATT TGCGACTGTT CCTGTCTTCC CGGATCGCCC TTCCCTTGTC CTCATTCTGG
                ◄──────────────────────────────────┘3' PRIMER
```

3. Working directly downstream from the repeat, locate a GC-rich segment without long stretches of the same base. Write the complementary base sequence of the bottom strand (or use the program at http://arep.med.harvard.edu/labgc/adnan/projects/Utilities/revcomp.html and paste the downstream sequence into the box and click "complement"). Copy an 18- to 30-nucleotide segment of this region and paste it into the text file. Rewrite the sequence to 5' to 3' (from 3' to 5'). This is a preliminary 3'-primer.
4. See Figure 9.1 for an example.

Part C: Evaluation of the Preliminary Designed Primers in IDT OligoAnalyzer 3.1

1. Access www.idtdna.com/Home/Home.aspx.
2. Select OligoAnalyzer.
3. Paste the preliminary 5' primer DNA sequence to the box.
4. Click Analyze. Copy and paste the results to the text file, including the GC-content, length, and melting temperature.
5. Now click Hairpin. Use the default values in the boxes and select submit. Save your results in the text file for the top hairpin.
6. Click Self-Dimer and submit. Save your results (first potential dimer) in the text file.
7. Repeat steps 3 through 6 for the 3' preliminary primer.
8. Evaluate the results. The primers should meet all of the criteria in Table 9.2. If they do not, design new primers by shifting the potential sequence up or downstream a few bases.
9. Evaluate Hetero-dimers by pasting the 5' primer sequence to the text box and click Hetero-dimer, add the 3' primer sequence to the second text box, and submit. Save your results (first potential hetero-dimer) in the text file.
10. If the primers do not meet the criteria set forth in Table 9.2, adjust the primers and reevaluate until you find a better pair. Save your results.

Part D: Using NCBI Primer-Blast to Search for Nonspecific Priming

1. Submit the primer pair to NCBI Primer-Blast (*Homo sapiens*) to evaluate if there is any complementarity at an unintended locus.
2. Go to www.ncbi.nlm.nih.gov/tools/primer-blast.

3. Enter the NCBI accession number in the PCR template box and the designed primers in the forward and reverse primer boxes under primer parameters. Indicate the primer region for the NCBI accession number by base number in the boxes to the right of the PCR template box and the expected product size below the primer sequence boxes. Blast the primers to the entire human genome sequence.
4. If the primers bind to more than one site in the genome, revise the primers as you did previously to be specific to one site. Save your results to the text file.

Part E: Creating a Multiplex of Two Primer Sets for Two Unique Sites

1. With another STR locus, tests a multiplex of two primer sets for two unique sites to be used together in a PCR reaction. Multiplexing with another person or lab group saves time.

TABLE 9.3   Student Data Template for Multiplex PCR Primer Design

Data Table	5′ Primer	3′ Primer
Locus 1		
STR Repeat		
NCBI Accession number		
Allele		
Primers		
Length (base pairs)		
GC content (%)		
T_M (°C)		
Self-Dimer (base pairs)		
Self-Dimer (kcal mol^{-1})		
Hairpin (base pairs)		
ΔG (kcal mol^{-1})		
T_M (°C)		
ΔH (kcal mol^{-1})		
ΔS (cal K^{-1}mol^{-1})		
Locus 2		
Primers		
Hetero-Dimer (base pairs)		
Hetero-Dimer (kcal mol^{-1})		

Check the multiplex primer pairs against each other in OligoAnalyzer using the Hetero-Dimer function ($5'/5'$, $5'/3'$, $3'/5'$, $3'/3'$ for the four primers of the two loci).

2. Edit your primers to create the best multiplex using the remaining lab time while recording all attempts in a data file (Table 9.3).

QUESTIONS

Report the data contained in the chart for the final iteration of your primers in IDT's OligoAnalyzer and NCBI's Primer-Blast. Provide a narrative of what you changed each time the primers were revised and explain why. Report your final primers. Report your results of multiplexing and the NCBI Accession number, locus, and primers for the group/individual used to multiplex the primers you designed. State if any features of the primers do not meet the desired criteria.

Compute the amplicon size for K562 (or other desired standard for subsequent testing) using www.dna-fingerprint.com/index.php?module=pagesetter&tid=1&orderby=locus, allele&filter=template:like:K562 to determine the number of STR repeats for the locus in that cell line.

References

BLAST, www.ncbi.nlm.nih.gov/blast, accessed 4/1/2012.

Butler, J.M., 2005. Forensic DNA Typing, second ed. Elsevier, Burlington, MA, pp 1—660.

Elkins, K.M., 2011. Designing Polymerase Chain Reaction (PCR) Primer Multiplexes in the Forensic Laboratory. Journal of Chemical Education 88, 1422—1427.

Kim, T.D., 2000. PCR primer design: an inquiry-based introduction to bioinformatics on the World Wide Web. Biochemistry and Molecular Biology Education 28, 274—276.

Lima, A.O.S., Garcês, S.P.S., 2006. Intrageneric primer design: Bringing bioinformatics tools to the class. Biochemistry and Molecular Biology Education 34, 332—337.

Masibay, A., Mozer, T.J., Sprecher, C., 2000. Promega Corporation reveals primer sequences in its testing kits [letter]. J. Forensic Sci. 45, 1360—1362. www.cstl.nist.gov/strbase/promega_primers.htm (accessed June 22, 2010).

Mitsuhashi, M., 1996. Technical report: Part 2; Basic requirements for designing optimal PCR primers. J. Clin. Lab. Anal. 10, 285—293.

Vallone, P.M., Butler, J.M., 2004. AutoDimer: A screening tool for primer-dimer and hairpin structures. Biotechniques 37, 226—231.

10

Testing Designed Polymerase Chain Reaction (PCR) Primers in Multiplex Reactions

OBJECTIVE

To evaluate the ability of the designed PCR primers to amplify the STR locus of interest in single and multiplex PCR reactions.

SAFETY

Wear gloves. Handle the SYBR Green I reaction mix carefully. SYBR Green I is an intercalating agent. Immediately wash off any chemicals with which you come into skin contact using soap and water.

The real-time PCR instrument is very sensitive to background noise. Do not touch the plate without gloved hands. To avoid contamination, always wear gloves when handling the samples and reagents.

MATERIALS

1. BioRad iQ 96-Well PCR Plates
2. BioRad Microseal "B" film
3. BioRad 2x iQ SYBR Green SuperMix
4. Nuclease-free water
5. K562 DNA template (diluted to 1 ng)
6. Designed and purchased Forward (5′) and Reverse (3′) PCR primers (diluted to 5 μM) (from 25 nmole, standard desalting)
7. BioRad iQ5 RT-PCR instrument with BioRad iQ5 software v. 2.0 or equivalent
8. Micropipettes (e.g., 0.5 to 10 μL, 1 to 20 μL, 10 to 200 μL, and 100 to 1000 μL) and autoclaved tips

9. Sterile 1.5-mL microcentrifuge tubes
10. Gloves
11. Freezer
12. Microwave or hot plate
13. Agarose
14. Gel box and 14-well comb and power supply
15. 100X SYBR
16. DNA ladder
17. 6X gel loading dye
18. Photodocumentation system with SYBR Green I filter

BACKGROUND

PCR primers often require optimization for the best results. This optimization can require significant time in evaluating varying concentrations for DNA polymerase, 2′-deoxynucleotide triphosphates (dNTPs), magnesium, KCl, Tris buffer, bovine serum albumin, and the designed and purchased PCR primers. The annealing temperatures of the primers also need to be evaluated experimentally for the best results. In this experiment, the annealing temperatures will be evaluated using gradient real-time PCR (Higuchi et al. 1993), but the concentrations of the other reagents will be fixed as in the purchased proprietary 2X reagent mixture containing dNTPs, 50 U/µL antibody-mediated hot-start *iTaq* DNA polymerase, 6 mM MgCl$_2$, SYBR Green I, enhancers, stabilizers, and 20 nM fluorescein. A 25 µL reaction volume will be used for all reactions.

As an example, for the TPOX STR PCR primers in Figure 10.1, the expected amplicon sizes for K562 are 100 and 104 bp (heterozygote alleles 8,9). A sample amplification curve derived from a PCR reaction with these primers is shown in Figure 10.2.

The expected melting temperature is 75.9° C/88.6° C/77.91° C as computed using the web tools at www.basic.northwestern.edu/biotools/oligocalc.html#helpbasic. The observed melting temperature is 79.5° C (Figure 10.3).

An agarose gel will be used to confirm the production of the expected PCR product(s). A 2% agarose gel can be used to compare the migration of the amplicon with that of the DNA size fragments in a DNA ladder (Figure 10.4).

In this experiment, the designed CODIS STR locus primers will be employed to amplify K562 (or another DNA standard) using gradient PCR to determine the optimal annealing temperature of the primers and if the correct amplicon is produced (Elkins and Kadunc, 2012).

```
                                          ⟹5′ PRIMER
GTTTCAGGGC TGTGATCACT AGCACCCAGA ACCGTCGACT GGCACAGAAC AGGCACTTAG
CAAAGTCCCG ACACTAGTGA TCGTGGGTCT TGGCAGGTGA CCGTGTCTTG TCCGTGAATC

STR REPEATS  1    2    3    4    5    6    7    8    9
GGAACCCTCA CTGAATGAAT GAATGAATGA ATGAATGAAT GAATGAATG TTT
CCTTGGGAGT GACTTACTTA CTTACTTACT TACTTACTTA CTTACTTAC AAA

GGGCAAATAA ACGCTGACAA GGACAGAAGG G
CCCGTTTATT TGCGACTGTT CCTGTCTTCC C
             3′ PRIMER           ⟸
```

FIGURE 10.1 PCR primer locations for TPOX primers used to amplify *Homo sapiens* thyroid peroxidase (TPO) on chromosome 2; NCBI Nucleotide Sequence (*Accession* M68651): repeat (gray), and sample-designed primers (arrows and bold).

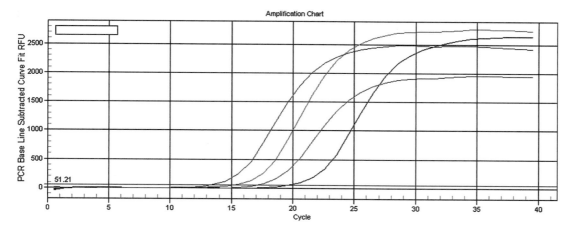

FIGURE 10.2 Sample exponential amplification curves that show the results of 40 cycles of gradient real-time PCR using primers depicted in Figure 10.1. Curves of amplified K562 DNA (from 1 ng starting material) as detected by SYBR Green I at annealing temperatures ranging from 50–65° C.

PROCEDURE

1. To each tube of purchased PCR primer, add 100 μL of nuclease-free water. Transfer 1 μL to a new 1.5-mL microcentrifuge tube, and add 999 μL of water. Quantify the DNA in the 1-mL sample by recording the absorbance at 260 nm using a UV-Vis spectrophotometer. Calculate the quantity of the diluted and undiluted samples using the extinction coefficient provided on the product insert using Beer's law. Dilute the original stock of PCR primer to 5 μM prior to use.

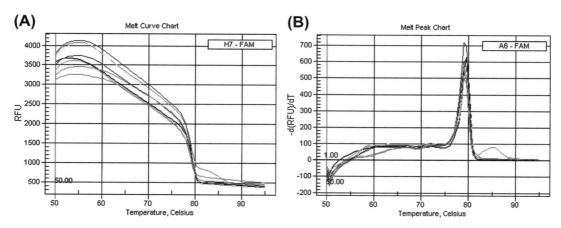

FIGURE 10.3 A. Melting curve plot from 50° to 95° C showing the temperature inflection at 79.5° C for the K562 DNA (left). B. First derivative plot of the melting curve from 50° to 95° C showing the predominant peak at 79.5° C for the K562 DNA (right) using the primers depicted in Figure 10.1.

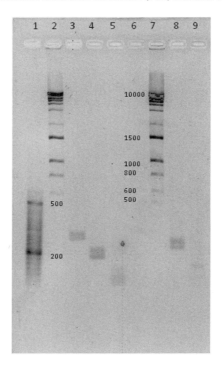

FIGURE 10.4 Sample 2% agarose gel of PCR of K562 DNA using designed primers. Lanes 1 and 7 contain the DNA Logic Ladder (sizes indicated), lane 2 contains a low molecular weight ladder (500 - 20 bp) (sizes indicated), and lanes 3 to 6 and 8 to 9 contain amplicons produced using experimental designed PCR primers. Lane 5 contains the amplicon produced from the primers shown in Figure 10.1.

2. Pipet the following identical samples into eight separate 1.5-mL microcentrifuge tubes for the standard DNA samples to prepare for gradient PCR (annealing temperatures to range from 50° to 65° C on the plate):
   12.5 μL of 2x iQ SYBR Green SuperMix
   1 μL of 5 μM designed forward primer
   1 μL of 5 μM designed reverse primer
   9.5 μL of nuclease-free water
   1 μL of K562 DNA (1 ng/μL)
   25 μL Total Reaction Volume
3. Prepare at least one negative control sample (substitute nuclease-free water for template DNA, so add 10.5 μL of nuclease-free water) for the plate.
4. Prepare one set of multiplex samples (substitute two paired forward and reverse primers for 2 uL of nuclease-free water and decrease water to 7.5 μL).
5. Mix samples by pipetting up and down gently so as not to introduce bubbles. Centrifuge samples briefly.
6. Pipette your samples into the 96-well plate by pipetting down the sides of the plate (or thin-walled PCR tube); carefully record wells used. Pipette sample into the bottom of the

well and do not release pipette until withdrawn from the sample so as to not produce bubbles. Load wells A to H for gradient PCR.

7. When all samples have been loaded to the 96-well plate, use the plastic squeegee device to seal Microseal plastic covers over all samples.

8. Open PCR machine for the instrument and insert well plate; fully close lid.

9. Turn on the computer attached to the BioRad iQ5 RT-PCR instrument. Turn on the BioRad iQ5 RT-PCR instrument; wait for the self-test to complete (approximately 10 minutes prior to use).

10. When the User Log-on screen appears, open the iQ5 v.2.0 software on the computer. Under the Setup tab, select Protocol and select Create New to set up a new experiment. In the table at the bottom of the screen, edit the cycle settings to read: Cycle 1, Step 1, 3:00 minute Dwell time, 95.0 Setpoint. After Cycle 1 is created, click on the ... under Options on the Cycle 1 line to insert cycles "After" and select "2 Steps" and select insert. Edit Cycle 2, Step 1 to read: 0.15 Dwell time at 95.0 Setpoint and Cycle 2: Step 2 to read: 1:00 Dwell time and 50 Real time Select Gradient Range 15 Setpoint (Figure 10.5). After Cycle 2 is created, click on the ... under Options on the Cycle 2 line to insert cycles "After" and select "1 Step" and select insert. In the drop-down box under PCR/Melt Data Acquisition, select Melt Curve. Edit the preset conditions to read: 50.0 Setpoint. This will execute a melting curve of 91 cycles for 30 seconds each of melting every 0.5° C from 50° C. Select Save and Exit Protocol Editing, give it a filename, and click save.

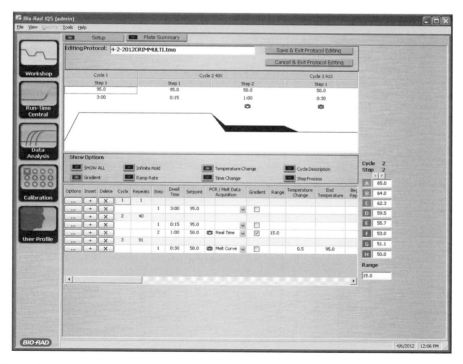

FIGURE 10.5 Setup of the gradient PCR amplification cycle and plate temperature protocol on BioRad iQ5.

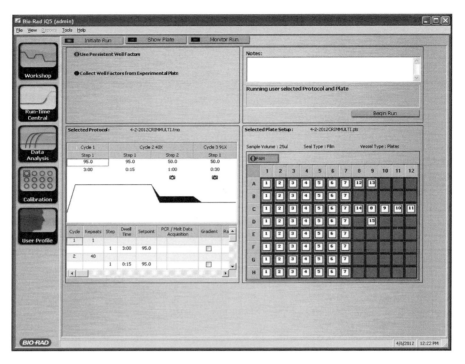

FIGURE 10.6   Setup of the gradient PCR run protocol on BioRad iQ5.

11. From the Setup page, now select Plate. Create New to input student sample information. Click on the box with the X in it to identify the student samples (all of these are performed with a positive control but at different annealing temperatures) and click in the column and row desired (e.g., A1 through H1) to the student sample with the designed primers for the gradient PCR. Clicking on the boxes highlights the sample for editing in the table below, and student sample identifiers can be added. Click on the negative control symbol and click in the column and row desired (e.g., A12 to H12) to insert a negative control sample. When completed, select Save and Exit Plate Editing and give the plate a filename (Figure 10.6).

12. In the top right-hand corner of the screen, select Run. The instrument was previously calibrated with the five dyes, so we select Use Persistent Well Factors for dye deconvolution. Select Begin Run and give it a data filename. The selected protocol and plate will have the filenames displayed to ensure proper run parameters.

13. When the run is completed, click OK in the box on the screen. The run saves automatically to the preset file.

14. To analyze the data, open the Data File. The first screen is the PCR Quant with all samples plotted for RFU versus cycle number. Clicking on threshold cycle yields the entrance to log phase for the setting given (move the green line to adjust). By clicking Display Wells, wells may be turned off or on for viewing by selecting the well to turn off and clicking OK.

15. The melt data can be analyzed in the Melt Curve Peak window. The left-hand plot is the RFU versus temperature (°C). The right-hand plot is the first derivative of the RFU versus temperature (°C). The RFU values and melt temperatures for the peaks are given in the table at the bottom of the screen.
16. The graphs can be exported using PDF Converter. Alternatively, the graphs can be copied to Paint for saving by right-clicking in the graph, selecting copy graph, and copying to Paint. Files can then be saved as .JPG or .TIFF files.
17. Optional second week. Prepare a 2% agarose gel (Refer to Experiment 5) using 1 g of agarose melted in 50 mL of 1X TAE buffer and set in a gel box with a 14-well comb. Add 3 μL of the 6X loading dye per 25 μL sample. Combine 8 μL of the DNA Logic ladder, low molecular weight ladder or preferred ladder with 1 μL of 100x SYBR Green I dye and 1 μL of the 6x loading dye. Load 12 μL samples and all of ladders while carefully RECORDING LOCATIONS of samples. Run at 150 V for a 10 cm gel for 60 minutes. Remove gel from gel box and record image with photodocumentation unit.

QUESTIONS

1. What were the expected and observed T_m values for the amplicons? Give reasons for any deviations.
2. Determine the best annealing temperature for the designed primers based on the gradient temperature that produced the most correct amplicon.
3. From the gel, estimate the size of the amplicon produced by measuring the distances of the ladder fragments and the amplicon fragment(s), plotting log bp versus distance for the ladder, and using the line equation to compute the size of the amplicon.
4. Comment on the production of the expected amplicons in the multiplexes.

References

Elkins, K.M., Kadunc, R.E., 2012. An Undergraduate Laboratory Experiment for Upper-Level Forensic Science, Biochemistry, or Molecular Biology Courses: Human DNA Amplification Using STR Single Locus Primers by Real-Time PCR with SYBR Green Detection. Journal of Chemical Education 89, 784–790.

Higuchi, R., Fockler, C., Dollinger, G., Watson, R., 1993. Kinetic PCR analysis: real-time monitoring of DNA amplification reactions. Biotechnology (NY) 11, 1026–1030.

11

Multiplex Polymerase Chain Reaction (PCR) Amplification of Short Tandem Repeat (STR) Loci Using a Commercial Kit

OBJECTIVE

To learn how to set up PCR reactions for DNA amplification using a commercial multiplex kit for amplifying autosomal nuclear DNA.

SAFETY

Handle all reagents with ethanol-rinsed gloves to avoid contaminating the kit components. The PCR-primers are labeled with fluorescent dyes; avoid contact with your hands.

MATERIALS

1. Extracted DNA simulated evidence template (1- to 2.5-ng range)
2. Applied Biosystems commercial kit (e.g., SGM Plus, Profiler Plus, COfiler, Identifiler, or other desired kit) complete with AmpliTaq Gold DNA Polymerase, PCR Reaction mix, and Primer set) (or Promega PowerPlex 16 kit or other desired kit) (a list of selected kits provided in Table 11.1)
3. Standard DNA (K562, 9947A, 007, 9948, SRM 2392, etc.)
4. PCR thermal cycler
5. Micropipettes and sterile tips (filter tips preferred) (0.5 to 10 μL, 2 to 20 μL, 10 to 100 μL, 100 to 1000 μL)
6. Nuclease-free water
7. Microcentrifuges
8. Microcentrifuge tube rack

9. Thin-walled PCR tube rack
10. Thin-walled PCR tubes
11. Disposable gloves
12. Thin line permanent marker
13. Vortexer

BACKGROUND

The DNA sequences of interest to forensic investigators are noncoding regions that contain segments of short tandem repeats (STRs). In the United States, these STRs are predominantly the CODIS STR loci. These short DNA sequences are repeated in a head-to-tail fashion (for example, an allele 4 or 4 tetranucleotide repeats is AATG-AATG-AATG-AATG, whereas an allele 3 or 3 tetranucleotide repeats is AATG-AATG-AATG). The result is differences in length or length polymorphisms. The analysis of short-tandem repeat loci is a length-based DNA typing system for human identification. A majority of the STRs that have been evaluated by the forensic community are composed of four-nucleotide repeat units, referred to as tetranucleotide STRs. Other DNA typing methods evaluate sequence, as opposed to length, polymorphisms. STRs are used for forensic, paternity, and missing persons casework, genealogical studies and archeological investigations. The sequences that differ are often polymorphic (of many forms). Humans share 99.5% identity in their DNA, so only a small percentage of the DNA differs (0.5%).

STR markers are polymorphic DNA loci that contain a repeated nucleotide sequence. The STR repeat unit can be from two to seven nucleotides in length. The number of nucleotides per repeat unit is the same for a majority of repeats within an STR locus. The number of repeat units at an STR locus may differ, so alleles of many different lengths are possible. Polymorphic STR loci are therefore very useful for human identification purposes. STR loci can be amplified using the PCR process, and the PCR products are then analyzed by capillary electrophoresis in a subsequent experiment to separate the alleles according to size. PCR-amplified STR alleles can be detected using various methods, such as radioactive nucleotide labeling and silver staining, but PCR products in modern kits are detected by fluorescent dyes attached to the 5' end of one primer in each set.

There are numerous PCR-based multiplex kits commercially available for typing human STRs. The kits are primarily focused on autosomal loci on one of the nuclear chromosomes 1 to 22. However, kits are also available to amplify regions on X- and Y-chromosomes (Applied Biosystems AmpFlSTR Yfiler, PCR Amplification Kit, User's manual, Mulero et al., 2006), and the autosomal DNA typing kits contain primers to amplify a locus such as amelogenin used to determine the gender of the individual who is the source of the DNA being probed. The loci are named according to the following system: the first letter is *D* for DNA, the first number is the chromosome number on which the locus is located, the second letter is *S* for single copy, and the second number corresponds to a location of the STR on the chromosome.

The PCR STR multiplex kits range in multilocus detection capability from 3 loci to 18 loci that can be probed in a single test tube reaction (Table 11.1). The AmpF*l* STR SGM Plus PCR Amplification Kit contains all the reagents necessary to amplify 10 STRs including the D3S1358, vWA, D16S539, D2S1338, D8S1179, D21S11, D18S51, D19S433, TH01, and FGA loci and amelogenin

TABLE 11.1 STR Loci Probed by Commercial Multiplex Kits

STR locus	Life Technologies - Applied Biosystems								Promega					
	Blue	Green 1	SGM Plus	Profiler	Profiler Plus	COfiler	Identifiler	Identifiler Plus	Powerplex 1.1/1.2	Powerplex 2.1/2.2	Powerplex 16.2	CTTV	FFFL	Gamma STR
D16S539			×			×	×	×	×		×			×
D7S820				×	×	×	×	×	×		×			×
D13S317				×	×		×	×	×		×			×
D5S818				×	×		×	×	×		×			×
CSF1PO		×		×		×	×	×	×		×	×		
TPOX		×		×		×	×	×	×	×	×	×		
AMEL		×	×	×	×	×	×	×		×	×			
THO1		×	×	×		×	×	×	×	×	×	×		
VWA	×		×	×	×		×	×	×	×	×	×		
D18S51			×		×		×	×		×	×			
D21S11			×		×		×	×		×	×			
D3S1358	×		×	×	×	×	×	×		×	×			
FGA	×		×	×	×		×	×		×	×			
D8S1179			×		×		×	×		×	×			
Penta D											×			
Penta E										×	×			
D2S1338			×				×	×						
D19S433			×				×	×						
F13A01													×	
FESFPS													×	
F13B													×	
LPL													×	

from sample DNA all in one test tube (Applied Biosystems AmpFlSTR SGM Plus, PCR Amplification Kit, User's manual). The AmpFl STR Profiler Plus PCR Amplification Kit contains all the reagents necessary to amplify nine STRs, including the D3S1358, vWA, FGA, D8S1179, D21S11, D18S51, D5S818, D13S317, and D7S820 loci and amelogenin from sample DNA all in one test tube (Applied Biosystems AmpFlSTR Profiler Plus, PCR Amplification Kit, User's manual). The AmpFl STR COfiler PCR Amplification Kit contains all the reagents necessary to amplify six STRs, including the D3S1358, D7S820, TH01, TPOX, CSF1PO, and D16S539 loci and amelogenin from sample DNA all in one test tube (Applied Biosystems AmpFlSTR COfiler, PCR Amplification Kit, User's manual).

INTERPOL, the European police network, has adopted a set of four STR loci, including FGA, D21S11, TH01, and vWA. The European Network of Forensic Science Institutes (ENFSI) has recommended seven STR loci, including FGA, D21S11, TH01, vWA, D8S1179, D18S51, and D3S1358. The Grupo Iberoamericano de Trabajo en Análisis de DNA (GITAD) has recommended six loci, including CSF1PO, TH01, TPOX, D16S539, D7S820, and D13S317 for STR DNA typing.

For the PCR reaction, the kit reagents contain deoxynucleotide triphosphates (dNTPs), including dATP, dTTP, dGTP, and dCTP, buffer, AmpliTaq Gold DNA polymerase, fluorescently labeled 5' and 3' primers for each locus in the multiplex, 0.05% sodium azide, bovine serum albumin, magnesium chloride and Tris buffer. The 5' primer for each locus is labeled with one of three fluorescent dyes with a unique excitation and emission wavelength. Multicomponent analysis and diffraction gratings are used to differentiate the dyes even if the colors exhibit spectral overlap. The red dye (ROX) is used to label the size standard. The reaction requires the analyst to add only the extracted DNA. The kit also contains an allele ladder for post-electrophoresis allele assignment. All reagents should be kept on ice throughout the duration of the lab, and the PCR reaction should be started immediately after adding all reagents. In the thermal cycler does not have a heated top; the samples and controls are topped with mineral oil to prevent evaporation. To run the PCR reaction, the tube must be inserted into a thermal cycler, a machine with a variable heating block. This lab utilizes a three-step cycle (the following numbers are the same for the SGM Plus/ProFiler Plus/ Cofiler/Identifiler kits): The AmpliTaq Gold polymerase is initially denatured at 95° C for 11 minutes. Then the following set of steps is repeated 28 times. (1) The DNA is denatured at 94° C for 1 minute. (2) The primers are allowed to anneal (base pair) at 59° C for 60 seconds. (3) The AmpliTaq Gold DNA polymerase attaches where the primer and DNA are bound and extends the primer at 72° C for 60 seconds. A final extension of 45 minutes is run at 60° C to finish the elongation of all strands including polyA+ addition (addition of A) to all strands to reduce split peaks on the capillary electrophoresis (in which one amplicon has an extra A and one does not and they differ by one base pair). As the autosomes are diploid, the STR loci will typically be a diplotype. The thermocycler can be set to refrigerate the samples (4° C) until they can be removed. After PCR, 2^{28} copies of one DNA template are made!

The application of PCR-based typing for forensic or paternity casework requires validation studies and quality-control measures. The quality of the purified DNA sample, as well as small changes in buffers, ionic strength, primer concentrations, choice of thermal cycler, and thermal cycling conditions, can affect the success of PCR amplification. STR analysis is subject to contamination by very small amounts of nontemplate human DNA. Extreme care should be taken to avoid cross-contamination in preparing sample DNA, handling primer pairs, setting up amplification reactions, and analyzing amplification products in different areas of the lab. Reagents and materials used prior to amplification (e.g., STR 10X Buffer, K562 control (or other standard) DNA, and dye-labeled 10X primer pairs) should be stored separately from those used following amplification (e.g., dye-labeled allelic ladders, loading solutions, and Gel Tracking Dye). Always include a negative no template control reaction to ensure reagent purity. Some of the reagents used in the analysis of STR products are potentially hazardous and should be handled accordingly.

Various samples will be run using the commercial multiplex kit. These include the evidence sample or sample collected from the crime scene, reference samples which may include substrate control samples (e.g., cutting of jeans from crime scene without blood or body fluid), samples from a victim or suspect or relatives, standard sample purchased and known to be a high quality DNA template and known profile, a positive control sample of known profile, and negative control. The negative control may include a no template control in which water is substituted for DNA in a reaction, or a reaction in which female DNA is substituted for male DNA when a male-specific reaction is performed.

In this experiment, PCR reactions will be set up for STR loci using a commercial multiplex kit such as the AmpFlSTR SGM Plus, Profiler Plus, COfiler, or Identifiler kits (Applied Biosystems AmpFlSTR Identifiler, PCR Amplification Kit, User's manual, Collins et al., 2004) from Applied Biosystems or another selected kit from another manufacturer.

PROCEDURE

Part A: Preparation of the DNA Template for PCR

Dilute previously extracted genomic DNA from known reference or simulated crime scene samples to 1 to 2.5 ng in 2 μL.

Part B: Preparation of the Positive Control Template (Known Template—K562 or Other)

The control reaction is set up by preparing DNA in the same manner as that described in Part A. The only difference is that 2 μL of K562 or other standard DNA template (1 to 2.5 ng) is added to the tube instead of the previously isolated genomic DNA.

Part C: Preparation of the Negative Control:

The negative control is prepared by substituting 2 μL of sterile distilled, deionized water to the reaction tube so as to maintain the same volume as the previous two reaction tubes.

Part D: Preparation of the PCR Mixture Using a Commercial Multiplex Kit

Vortex the AmpFl STR PCR Reaction Mix, AmpFl STR SGM Plus or ProFiler Plus (or other) Primer Set, and AmpliTaq Gold DNA Polymerase for 5 seconds. Spin the tubes briefly in a microcentrifuge to remove any liquid from the caps.

Compute the number of total reactions to run and divide the number of reactions by half and add one to prepare enough master mix as shown next. For example, for eight reactions, prepare four and a half times the quantity of the reagents for the master mix.

Master Mix per Two Half Reactions

AmpFl STR PCR Reaction Mix	21.0 µL
AmpFl STR SGM Plus Primer Set	11.00 µL
AmpliTaq Gold DNA Polymerase (5µ/uL)	1.0 µL
	33.00 µL total volume

Component Volume (Half Reactions)

AmpFl STR PCR ProFiler Plus Reaction Mix	10.5 µL
AmpFl STR ProFiler Plus Primer Set	5.50 µL
AmpliTaq Gold DNA Polymerase (5u/µL)	0.5 µL
	16.50 µL total volume

Sample

Pipette 2 µL (1 to 2.5 ng DNA) of the previously isolated genomic DNA from Part A into the tube containing 16.5 µL of master mix and 6.5 µL of nuclease-free water.

Positive Control

Add 2 µL (2 ng) of AmpFl STR control DNA 007 or K562 control DNA from Part B to the labeled Positive Control Tube containing 16.5 µL of master mix and 6.5 µL of nuclease-free water.

Negative Control

Add 8.5 µL of sterile water from Part C to the labeled Negative Control Tube containing 16.5 µL of PCR mix.
Mix thoroughly by vortexing at medium speed for 5 seconds.
Spin the tube briefly in a microcentrifuge to remove any liquid from the cap.
If the thermocycler does not have a heated cap, add one drop of mineral oil.

Part E: The Temperature Cycle Used for PCR Amplification of the 11 STR Loci (SGM Plus, Profiler Plus, COfiler or Identifiler)

Place all the tubes in the PCR Thermal Cycler, and begin the preprogrammed cycles listed as follows:

1. 95° C for 11 minutes
2. 94° C for 60 seconds

59° C for 60 seconds
72° C for 60 seconds
Repeat the above cycle 28 times.

3. 60° C for 45 minutes to finish the elongation of all strands.

4. Hold at 25° C until the next morning. Your instructor will remove your samples and store them in the freezer until next week.

QUESTIONS

Define and compare the use of evidence, reference, standard, and positive and negative control samples. State what happens at each of the temperatures used in PCR. Predict the expected DNA profile of the standards using dna-fingerprint.com.

References

Applied Biosystems AmpFlSTR COfiler, PCR Amplification kit, User's manual.

Applied Biosystems AmpFlSTR Profiler Plus, PCR Amplification kit, User's manual.

Applied Biosystems AmpFlSTR SGM Plus, PCR Amplification kit, User's manual.

Applied Biosystems AmpFlSTR® Yfiler™ PCR Amplification Kit User's Manual, http://www3.appliedbiosystems.com/cms/groups/applied_markets_support/documents/generaldocuments/cms_041477.pdf. (accessed 11-22-11).

Collins, P.J., Hennessy, L.K., Leibelt, C.S., Roby, R.K., Reeder, D.J., Foxall, P.A. 2004. Developmental Validation of a Single-Tube Amplification of the 13 CODIS STR Loci, D2S1338, D19S433, and Amelogenin: The AmpFlSTR® Identifiler® PCR Amplification Kit. J. Forensic Sci. 49, 1265–77.

Genotypes for standards, DNA Fingerprint: http://www.dna-fingerprint.com.

Mulero, J.J., Chang, C.W., Calandro, L.M., Green, R.L., Li, Y., Johnson, C.L., Hennessy, L.K., 2006. Development and validation of the AmpFlSTR Yfiler PCR amplification kit: a male specific, single amplification 17 Y-STR multiplex system. J. Forensic Sci. 51, 64–75.

Capillary Electrophoresis of Short Tandem Repeat (STR) Polymerase Chain Reaction (PCR) Products from a Commercial Multiplex Kit

OBJECTIVE

To use capillary gel electrophoresis to separate multiplex STR PCR products amplified using a commercial kit and analyze the resulting electropherogram.

SAFETY

Caution: Extremely high voltage is used with the ABI 310 Genetic Analyzer and other capillary electrophoresis instruments. POP-4 is an acrylamide-based polymer. Although this material is polymerized, acrylamide is a neurotoxin and absorbed through the skin if any unpolymerized parts are in the mixture. Wear nitrile gloves.

MATERIALS

1. ABI 310 Genetic Analyzer
2. Alconox detergent
3. Formamide, deionized
4. Long Ranger, 50% stock solution
5. Matrix standards:
   a. Dye Primer Matrix Standard Kit (four-color) (Important: Do not use the TAMRA matrix standard provided in the kit.)
   b. NED Matrix Standard
6. 10X Genetic Analyzer buffer with EDTA
7. Pipet tips for gel loading, 0.2-mm flat tips

8. Micropipettes, adjustable, 0.5 to 10 μL, 2 to 20 μL, 20 to 200 μL, and 200 to 1000 μL, and filter tips
9. ABI PRISM 310 10X Genetic Analyzer Buffer with EDTA
10. G501 X8 ion exchange resin
11. GeneScan-500 (ROX) Size Standard Kit (includes Blue Detrax Loading Buffer)
12. Performance Optimized Polymer 4 (POP-4)
13. DNA Ladder for the kit used
14. 96-Well Plate Septa
15. Heating-cooling block or PCR instrument
16. POP-4 Performance Optimized Polymer
17. Autosampler 96-well Plate Kit
18. Polymer-reserve syringe, 5-mL glass syringe
19. 310 Capillaries, 47 cm × 50 μm i.d. (internally uncoated) (green)
20. 0.5-mL Sample Tray
21. Genetic Analyzer Retainer Clips (96-Tube Tray Septa Clips)
22. Genetic Analysis Sample Tubes (0.5 mL)
23. Genetic Analysis Septa for 0.5-mL Sample Tubes
24. Matrix Standard Set DS-33 (6FAM, VIC, NED, PET, and LIZ dyes) for 310/377 systems

BACKGROUND

The multiplex STR PCR products amplified using a commercial kit (Applied Biosystems AmpFlSTR COfiler, Identifiler, Profiler Plus or SGM Plus, PCR Amplification Kit, User's manual) will be separated using capillary electrophoresis (CE) in this experiment.

Multicomponent analysis is the process that separates the four different fluorescent dye colors into distinct spectral components using filters and matrices for color deconvolution. The three fluorescent dyes used in the AmpFl STR SGM Plus, COfiler, and Profiler Plus PCR Amplification Kits are 5-FAM, JOE, and NED (Applied Biosystems AmpFlSTR COfiler, Identifiler, ProFiler Plus and SGM Plus, PCR Amplification Kit, User's manuals). The fourth dye, ROX, is used for the GeneScan-500 Internal Lane Size Standard.

CE is a relative technique. Each sample—including the questioned samples, unknown samples, and allele ladder—is run with the internal standard. The known sizes of the size standard are plotted against the elution time and a line equation is used to size the unknown fragments using their elution times. The STR fragments are sized using the allele ladder consisting of DNA fragments corresponding to the sizes of the alleles at the probed loci using the elution times.

The Applied Biosystems (ABI) 310 instrument uses a 10-mW argon ion laser that excites the fluorescent dyes by the 488- and 514-nm lines. Each of the fluorescent dyes emits its maximum fluorescence at a different wavelength. During data collection on the Applied Biosystems (ABI) 310 single capillary instrument (Figure 12.1), the fluorescent signals are separated by a diffraction grating according to their wavelengths and projected onto a CCD camera in a predictably spaced pattern from 525 to 650 emission wavelengths. 5-FAM emits at the shortest wavelength and is detected as blue, followed by JOE (green), NED (yellow), and ROX (red), listed in order of increasing wavelength.

FIGURE 12.1 ABI 310 Genetic Analyzer, a single capillary electrophoresis instrument.

Although each of these dyes emits its maximum fluorescence at a different wavelength, there is some overlap in the emission spectra between the four dyes. The goal of multicomponent analysis is to isolate the signal from each dye so that, for example, signals from 5-FAM labeled products are displayed in the electropherogram for blue, but not in those for green, yellow, or red.

The output is an electropherogram, a chromatographic display with fluorescence intensity indicated as relative fluorescence units (RFU) on the y-axis and size on the x-axis based on the elution time matching between the sample and ladder with the same internal standard. After the internal lane size standard has been defined and applied, the electropherogram can be displayed with the base pair size on the x-axis.

The allelic ladders vary between kits and are specific for the amplicons and amplicon sizes produced by the primers in that kit (Applied Biosystems AmpFlSTR SGM Plus, PCR Amplification Kit, User's manual). For example, the AmpFlSTR SGM Plus allelic ladder contains the most common alleles for each locus. Genotypes are assigned by comparing the sizes obtained for the unknown samples with the sizes obtained for the alleles in the allelic ladder. Additional alleles have been included in the AmpFlSTR SGM Plus Allelic Ladder for the vWA, D21S11, D18S51, TH01, and FGA loci compared to those included for these same loci in other AmpFlSTR kits. Alleles contained in the AmpFlSTR SGM Plus Allelic Ladder and from population samples have been sequenced to verify both length and repeat structure. The size range is the actual base pair size of the sequenced alleles including the 3'

A+ nucleotide addition. The AmpFl STR SGM Plus PCR Amplification Kit is designed so that a majority of the PCR products contain the nontemplated 3' A+ nucleotide.

The alleles have been named in accordance with the recommendations of the DNA Commission of the International Society for Forensic Haemogenetics (ISFH). The number of complete 4-bp repeat units observed is designated by an integer. Variant alleles that contain a partial repeat are designated by a decimal followed by the number of bases in the partial repeat—for example, an FGA 26.2 allele contains 26 complete repeat units and a partial repeat unit of 2 bp.

In this laboratory, the PCR-amplified STR fragments for the loci amplified using a commercial multiplex kit will be separated on an ABI 310 Genetic Analyzer using a 47-cm capillary, POP-4 polymer, 24-minute run time, 15 kV injection voltage, and 5-second injection. The data will be analyzed using GeneMapper 4.0 or GeneMapperID in conjunction with the technical manuals for the kits, which indicate the expected fragment sizes for each allele.

PROCEDURE

Part A: Preparing Samples for Capillary Electrophoresis

1. Prepare DNA PCR samples using 12 µL of deionized formamide, 0.5 µL size standard (ROX), and 1 µL of PCR product (centrifuged before opening) for each sample and mix by vortexing.
2. Heat and denature each sample for 3 minutes at 95° C.
3. Snap-cool in an ice-water bath or freezer block for 30 seconds.
4. Load samples to the 96-well sequencing plate, and cover the samples with the rubber septa. (Note: Septa can be reused if soaked in bleach for 10 minutes.)

Part B: Setting up the Instrument

1. Wash the syringe with deionized (DI) water.
2. Rinse the 47-cm capillary for fragment analysis in DI water. The capillary may be used for 100 reactions and up to 200 to 300 reactions if it is stored properly in DI water when not in use.
3. Rinse the gel block with DI water, and loosen the screws. Do not loosen the o-rings.
4. Dilute 10X buffer with EDTA to 1X. Place 1X buffer in one vial and DI water in the other vial.
5. Assemble the gel block with the pins as shown, place on the CE, and add the buffer vial to the bottom of the block (Figure 12.2).
6. Clean the capillary window with ethanol using a Kimwipe to remove fingerprints or other trace compound.
7. Feed the capillary into the junction.
8. Place the capillary on the heating plate (Figure 12.3) so that the window block is in front of the laser and the capillary is lower than the electrode (metal wire).
9. In Autosampler calibration, follow the directions onscreen so that the capillary lines up with the dot (dot between the two) (Figure 12.4). When satisfied, select Set. Repeat for back dot in the same manner.

FIGURE 12.2 Assembling the gel block.

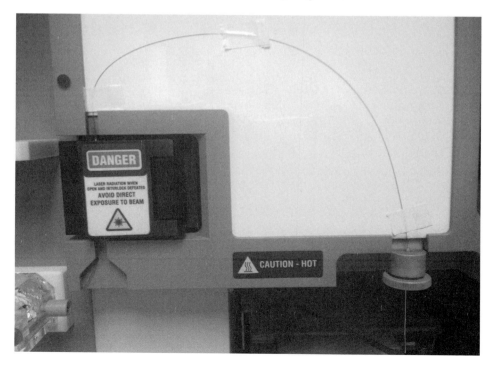

FIGURE 12.3 Inserting the capillary.

FIGURE 12.4   Aligning the autosampler.

10. Select Window > Manual Control > Autosample to position > Manual present tray. Fill the number 1 vial with 1X buffer with EDTA and vials numbers 2 and 3 with DI water (Figure 12.5).
11. Gently, draw some POP-4 up into the syringe, and rinse the sides of the syringe so water from the rinse step doesn't dilute polymer. Purge.
12. Carefully fill the syringe so as to avoid bubbles. Ensure that you load sufficient polymer for 300 uL to fill the gel block and 10 uL to run a reaction.
13. Screw in the loaded syringe. Select Manual control > Syringe down to engage.
14. Select Manual Control > Open buffer valve to fill polymer completely in gel block.
15. Select Close buffer valve >Execute.
16. Select Syringe Down > Use your thumbs to loosen the expunge valve. Retighten screws so the gel block does not leak polymer. Move the plate to position 1.
17. Select Autosampler > Present tray. Place the tray in autosampler gently.
18. Select Autosampler up.
19. Close heat plate and instrument doors.
20. Select Autosampler > Return tray.
21. Select File > New > Sample List > Add samples in GeneScan Injection sheet (Figure 12.6) using 4-dye matrix. Open GeneScan-Smpl Sheet 96 tube, Sample Sheets/DATE.

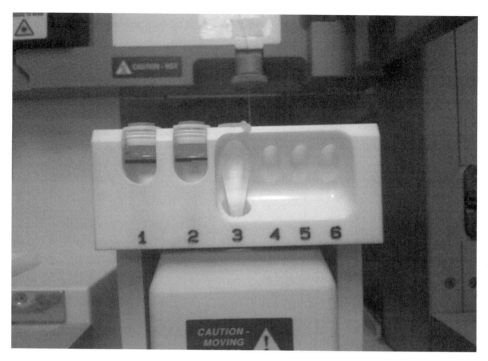

FIGURE 12.5   Filled buffer vials.

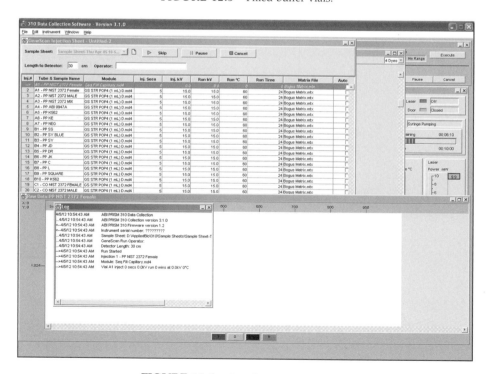

FIGURE 12.6   GeneScan Injection Sheet.

22. Select File > Save (give name). Select File > New > GeneScan Injection List. Set GeneScan injection-manual control-status to 60° C, Autosampler present tray-insert, Autosampler to position 1-exc. 4x to capillary in buffer. Autosampler up.
23. Select Instrument > Module > Reset capillary.
24. In GeneScan Injection Sheet, once samples are inputted, copy sample number 1 and insert at the top row and set: Sequence set capillary.
25. Run the experiment. Pumping the polymer will take approximately 1 hour, and each sample will require 24 minutes. Twenty minutes are required between samples to purge the used polymer and load the new polymer to the capillary.

Part C: Calling the Alleles and Analyzing the Data

1. Open GeneMapper 4.0 or GeneMapperID (Figure 12.7). Enter the password to open the file.
2. File: Open project-select one to view. Add sample to project.
3. Set the Analysis Method, Size Standard (ROX), Matrix, Panel (for kit used).
4. Select a file by clicking on it. Display plot as desired by toggling the color display on/off, loci label on/off, allele calls on/off, indicating number of panes per sheet, and so on.
5. Check all of the sample injections for anomalies, mixtures, and electrophoretic aberrations.

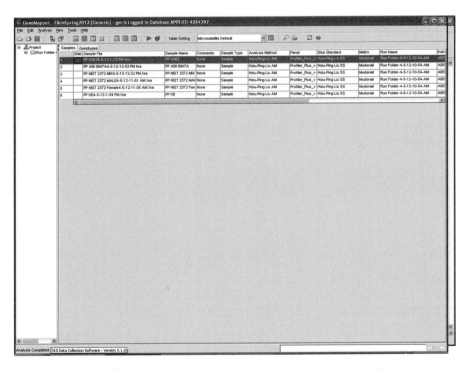

FIGURE 12.7   Using GeneMapper software to analyze results.

6. Check that all ladder alleles are correctly assigned.
7. Check the identity of all off-ladder alleles and confirm (if necessary).
8. Check that all controls (extraction, positive) are associated with the correct profile(s).
9. Check all reagent blanks and negative amplification blanks for anomalous peaks.
10. Check that all GS500 peaks are correctly assigned.
11. Check all peak height ratios (potential mixture).
12. Check for percentage stutter (potential mixture).
13. Check that identified mixtures and single source samples are unattributable to case samples compared against DNA profiles from DNA laboratory personnel and other samples in the batch.
14. For your data, note homo- or heterozygosity, possible mixture, polyploidism, pull-up, allele ladder, fragment size, allele call (by size), and stutter.

QUESTION

Assign your DNA electropherogram alleles for all loci for the kit you used (refer to the Applied Biosystems' or other kit manuals to check data by hand).

References

Applied Biosystems AmpFlSTR COfiler, PCR Amplification kit, User's manual.
Applied Biosystems AmpFlSTR Identifiler, PCR Amplification kit, User's manual.
Applied Biosystems AmpFlSTR ProFiler Plus, PCR Amplification kit, User's manual.
Applied Biosystems AmpFlSTR SGM Plus, PCR Amplification kit, User's manual.

13

Computing Random Match Probability from DNA Profile Data Using Population Databases

OBJECTIVE

To learn how to compute the probability of a random match and the discriminating power for STR multiplex DNA typing results using population databases.

SAFETY

No special safety precautions. This is a computational lab.

MATERIALS

1. Computer or calculator
2. STR genotype results
3. Population database: Budowle et al., 2001, Butler et al., 2003, or Figure 18.4 (Budowle et al., 1999)

BACKGROUND

Statistics is the science of collecting and analyzing large quantities of numerical data to compute the uncertainty of a result. The concept of probability is the numerical measure of the uncertainty of a given result. The term *probability of a random match in a population* is a statistical result. Population databases are based on probability, or the number of times an allele is observed divided by the total number of alleles for the locus in the population. If a DNA profile of a suspect does not match the profile produced from biological material obtained from the crime scene, then the suspect is not considered to be the donor of the biological evidence. However, if the DNA profile from the crime scene matches with a given suspect, it does not necessarily imply that a given individual is the exclusive donor of the material; indeed, it is important to correctly and precisely determine how common or rare the event is.

There are three possible outcomes as a result of comparing two DNA profiles. The first is inconclusive. An inconclusive result means that there are not enough data to determine a positive match. An exclusive outcome means that the DNA profiles of the biological evidence and the suspect donor are so different that they could not possibly have arisen from the same individual. An inclusive result means that the biological evidence matches the DNA profile of the suspect donor. These results are expressed as probabilities of random match of the same profile within the same population.

Probability values must range from zero to one. A zero means that an outcome is impossible or not counted, and a one means that an outcome is certain or that the event occurred. For independent events, the probability of an event occurring as computed from one event does not impact the probability of the same event occurring from a second event. Mathematically, the overall probability of the two events both occurring is the product of the probabilities of the independent events. Other events may be mutually exclusive: for one event to occur, another has to occur first. The overall probability of both of these events occurring is the sum of the probabilities of the two events.

In forensic science, allele probabilities are based on the evaluation of populations. Blood or buccal samples are solicited from a few hundred or more unrelated, anonymous, and consenting individuals from the population of interest (e.g., American-Caucasians, African-Americans, Asian-Americans). These are especially important for determining allele frequencies in regions/locations with isolated or inbred populations. Each of these individuals is profiled at each of the 13 CODIS (or other desired database) loci using a commercial multiplex kit. The results are tabulated and the number of occurrences of each allele is counted for each locus (Budowle et al., 2001, Butler et al., 2003). For STRs, alleles are numbered by the number of tandem repeats. The counts for each allele are divided by the number of possible alleles, or the number of individuals times two alleles for a diploid autosomal locus:

$$P = \text{Counts of allele \# observed}/2^*(\text{Number of individuals in the population})$$

This value is a probability of the event, or the allele, occurring. Each allele should be observed five times to be included in reliable statistical calculations. In cases where five observations are not made, the minimum allele frequency is computed by dividing 5 by 2N where N is the number of individuals in the population database and 2N is the number of chromosomes contained in those individuals:

$$P = 5/(2N)$$

If rare allele frequencies are less than the minimum allele frequency, replace the observed allele frequency with the minimum allele frequency.

We will not perform a population study in this experiment, but the data previously collected at the U. S. National Institute of Standards and Technology (NIST) will be used to calculate the rarity of the allele profiles assigned by CE (Table 18.4). The genotype is an inherited genetic variation. It may be observed as a single base polymorphism, allele for a short tandem repeat, or representation of a trait. The phenotype is a visible inherited trait or result of gene expression such as eye color, hair color, or temperament. The most common genotype can be computed using the most common alleles or those with the highest probability at all loci of interest. The least common genotype can be computed using the least common alleles or those with the lowest probability at all loci of interest.

Once the frequencies of alleles are computed for a given population, scientists can compare variation in allele frequencies between populations. Overall, the observation is that the genetic difference between individuals within the same racial groups is much greater than the difference of the average individuals compared of different races. Additionally, allele frequencies are not stagnant. Mutation, migration, natural selection, and random genetic drift all affect the probabilities of alleles observed in a population.

According to Mendel's laws of inheritance or Mendelian genetics, offspring inherit only one copy of a gene from each parent and the inherited allele is random. Indeed, Mendel's first law is the law of segregation. Hereditary traits are genes that occur in pairs, and during sexual development, these pairs are separated, or segregated, into sex gametes, or sex cells, and only one copy of the gene is passed to the offspring in the gamete as a result of meiosis. The alleles of different STR loci are inherited like any other Mendelian genetic markers. If an individual has two copies of allele 7 at locus TH01, the genotype is 7,7. For this to occur, both parents would have to have had a copy of allele 7 at TH01. Mendel's second law of inheritance is the law of independent assortment. This law states that two genes will assort independently and randomly and be inherited completely independent of each other; if a pair of genes are responsible for a trait, or phenotype, they will be independent of another pair of genes coding for another trait.

The allele frequencies are used to calculate the genotype frequency for each locus, assuming that the allele is following the Hardy-Weinberg equilibrium (HWE). The Hardy-Weinberg equilibrium states that the proportions of genotypes in a population are achieved in a single generation of mating, so p and q remain constant and allele frequencies remain constant from one generation to the next.

To perform statistics on a DNA profile, including that from the CE experiment, first download the table of allele frequencies (www.cstl.nist.gov/biotech/strbase/NISTpopdata/JFS2003IDresults.xls). Allele frequencies are probabilities used to compute genotype frequencies, or statistics.

The homozygous genotype frequency is simply defined as p^2. In Mendelian genetics, this is derived from an independent, biallelic locus with a dominant, p, and recessive, q, trait. Consider the example a trait for which the dominant allele is A and the recessive allele is a. A heterozygote cross yields a homozygous dominant (AA), two heterozygotes (Aa), and a homozygous recessive (aa) for the trait. In forensics with multiple STR alleles, a homozygote is typically referred to as p^2 (as opposed to q^2) because of the polymorphic loci used:

$$\text{Allele frequencies add to } 1 : p + q = 1$$
$$(p + q)^2 = p^2 + 2pq + q^2 = 1$$
$$\text{Genotype frequency or P (homozygote)} = p^2$$
$$\text{Genotype frequency or P (heterozygote)} = 2pq$$

In a homozygote, the probability of having two copies of allele 7 for TH01 or this genotype at the TH01 locus is determined by squaring the probability of having one allele, so $p^2 = 0.1940^2 = 0.03625$ for a Caucasian. The heterozygote genotype frequency is defined as $2pq$. Therefore, for a heterozygote, the probability of having one copy of allele 7 (p) and one copy of allele 8 (q) is 2(0.1940)(0.08444) = 0.03215. For each locus, the sum of all of the allele frequencies at a locus will add to a probability of 1.

Additionally, consider an individual with the genotype 15, 18 at the locus D3S1358. This individual is a heterozygote. In a reference database of 200 U.S. Caucasians, the frequency of the alleles 15 and 18 is 0.2825 and 0.1450, respectively. The frequency of the 15, 18 heterozygous genotype is therefore P = 2(0.2825)(0.1450) = 0.0819, or 8.2%.

Following determination of the genotype frequency, the frequency of the complete profile is estimated by multiplying the frequency of each locus (product rule). This assumes no genetic linkage between each locus and random assortment of alleles during mating. The overall profile probability can be determined by multiplying the probability of a genotype computed for each locus at each of the evaluated loci:

$$\text{Profile probability} = (P_1)(P_2)(P_3) \dots (P_n)$$
$$\text{Probability of a random match} = 1/(\text{Profile probability})$$

The resulting overall profile frequency for even 10 loci as normally evaluated by forensic scientists worldwide is very, very small, and the overall probability of a random match is determined by taking the reciprocal of the genotype frequency and may be reported as "there is a 1 in a 1,000,000,000 (or other computed number) chance that a particular unrelated person would have the same DNA profile."

However, there may be departures from the Hardy-Weinberg equilibrium. The assumptions of the Hardy-Weinberg equilibrium include a large population, absence of natural selection, mutation or migration, and random mating. Departures may be characterized by inbreeding or a high number of related individuals within a population, in which more homozygotes than expected will be observed; or if certain genotypes affect reproduction capability, in which some individuals have more offspring than expected and their alleles will be overrepresented; or if a genotyping error occurs and some genes are not represented because of a problem with the genotyping method.

The HWE may be used in testing for the independence of alleles within a locus. The genotype frequencies should match the allele frequencies, and the allele and genotype frequencies should be stable between generations. Excess homozygosity, allelic dropout, and null alleles are indications of deviations from HWE.

Corrections can be made for deviations from the HWE. When homozygous loci are present, the genotype frequency calculation requires a term called *theta* (θ), a correction factor needed because of the potential for nonrandom mating in a specific population (NRCII Recommendation 4.1), and the value of theta is usually considered to be approximately 0.01 to 0.03, depending upon racial group and isolation of the population. Typically, θ = 0.01 is used for the general population of United States, and θ = 0.03 is used for isolated/ inbred populations (Evett et al., 1996). Theta is needed because in an inbred population, the error caused by assuming HW equilibrium can overestimate the strength of the evidence and favor the prosecution in magnitude. In forensics, HWE is assumed, and the value of theta may or may not be used as a correction factor:

$$\text{Corrected genotype frequency or P (homozygote)} = p^2 + p(1-p) * \theta$$
$$\text{Corrected genotype frequency or P (heterozygote)} = 2pq * (1-\theta)$$

TABLE 13.1 Probability of a Random Match Calculation for 9947A Profile Using Allele Frequencies Provided on STRbase (Butler et al., 2003)

13 CODIS Loci		Allele Frequency (NIST Caucasian, N=302)			
Locus	9947A	Allele 1	Allele 2	p^2 or 2pq	$\Theta = 0.03$
Amelogenin	X, X	Female		n/a	
D3S1358	15, 14	0.26159	0.10265	0.05370	0.05209
D5S818	11, 11	0.36093	0.36093	0.13027	0.13719
vWA	18, 17	0.20033	0.28146	0.11277	0.10939
TH01	9.3, 8	0.36755	0.08444	0.06207	0.06021
D13S317	11, 11	0.33940	0.33940	0.11519	0.12192
D8S1179	13, 13	0.30464	0.30464	0.09281	0.09916
D21S11	30, 30	0.27815	0.27815	0.07737	0.08339
D7S820	11, 10	0.20695	0.24338	0.10073	0.09771
TPOX	8, 8	0.53477	0.53477	0.28598	0.29344
D16S539	12, 11	0.32616	0.32119	0.20952	0.20323
D18S51	19, 15	0.03808	0.15894	0.01210	0.01174
CSF1PO	12, 10	0.36093	0.21689	0.15656	0.15187
FGA	24, 23	0.13576	0.13411	0.03641	0.03532
		Profile Probability (Product of all loci)		0.2×10^{-13}	0.2×10^{-13}
		Probability of a Random Match (1 in ...)		5.9×10^{13}	5.7×10^{13}

In this experiment, the genotype frequency, profile probability, and probability of a random match will be computed for the STR profile obtained in Chapter 12 or a published profile. The probability calculations for 9947A standard DNA are shown in Table 13.1.

PROCEDURE

1. From the data collected and tabulated in the Capillary Electrophoresis experiment using a commercial multiplex kit, assign the allele frequency values from the database. State the racial group that you use for the allele frequency values obtained from Budowle et al., 2001, Butler et al., 2003, or Table 18.4.
2. Compute the genotype frequency for each locus.
3. Compute the profile probability.

4. Compute the probability of a random match.
5. Recompute these calculations using the correction factor 0.03 for loci.

QUESTIONS

Answer the questions in the procedure. What was the effect of the correction factor on the probability of the random match?

References

Budowle, B., Shea, B., Niezgoda, S., Chakraborty, R., 2001. CODIS STR Loci Data from 41 Sample Populations. J. Forensic Sci. 46, 453–489.

Butler, J.M., Schoske, R., Vallone, P.M., Redman, J.W., Kline, M.C., 2003. Allele Frequencies for 15 Autosomal STR Loci on U.S. Caucasian, African American, and Hispanic Populations. J. Forensic Sci. 48, 908–911.

Evett, I.W., Gill, P.D., Scrange, J.K., Weir, B.S., 1996. Establishing the Robustness of Short-Tandem-Repeat Statistics for Forensic Applications. Am. J. Hum. Genet. 58, 398–407.

Mitochondrial Deoxyribonucleic Acid (mtDNA) Single Nucleotide Polymorphism (SNP) Detection

OBJECTIVE

To learn how to set up and run real-time polymerase chain reactions (PCRs) to evaluate the base identity in a mitochondrial DNA single nucleotide polymorphism (SNP).

SAFETY

Wear gloves. Handle the SYBR Green I reaction mix carefully. SYBR Green I is an intercalating agent. Using soap and water, immediately wash off any chemicals with which you come into skin contact.

The real-time PCR instrument is very sensitive to background noise. Do not touch the plate without gloved hands. To avoid contamination, always wear gloves when handling the samples and reagents.

MATERIALS

1. BioRad iQ 96-Well PCR Plates
2. Microseal "B" film
3. 2x iQ SYBR Green SuperMix
4. Nuclease-free water
5. Extracted DNA template (1- to 10-ng range)
6. 16519F and 16519R primers (diluted to 5 μM) (IDT, custom, 25 nmole, standard desalting)
   Forward: 5'-ACCACCATCCTCCGTGAAAT-3'
   Reverse 1: 5'-CGTGTGGGCTATTTAGGCTTTACGA-3'
   Reverse 2: 5'- CGCGGCCGGCC-CGTGTGGGCTATTTAGGCTTTACGG-3' (with GC-rich clamp)

7. 10-ng/μL K562 DNA, 9947A, 9948, 007 DNA, simulated evidence DNA, etc. diluted to 1 ng/μL
8. BioRad iQ5 RT-PCR instrument with BioRad iQ5 software v. 2.0
9. Various variable volume micropipettes (e.g., 0.5 to 10 μL, 2 to 20 μL, 10 to 100 μL, 100 to 1000 μL) and autoclaved tips
10. 1.5 mL microcentrifuge tubes
11. Microcentrifuges
12. Vortex machines
13. Gloves
14. Freezer
15. Microwave or hot plate
16. Agarose
17. Gel box and power supply
18. 100X SYBR
19. DNA ladder
20. 6X gel loading dye
21. Photodocumentation system with SYBR Green I filter

BACKGROUND

SNPs are sequence polymorphisms identified in the human genome. Although SNPs vary in only one base, they are more common than short tandem repeats (STRs) and can yield valuable information. Approximately 85% of human variation is due to SNPs. SNPs have only two alleles and thus do not yield the discriminating power of an STR locus. In forensic science, SNPs have been used to assess the physical traits of suspects, including eye and hair color, as well as to provide an alternative and complementary typing method to STR DNA typing. SNPs have low mutation rates and may serve as suitable markers for lineage studies, missing persons cases, paternity cases, and cases where no direct reference sample is available. As the DNA databases are based on STR loci, it is unlikely, though, that SNPs will replace the older method.

There are many methods of detecting SNPs, including allele-specific hybridization, primer extension, allele-specific oligonucleotide ligation, DNA sequencing, PCR and restriction enzyme digestion, allele-specific oligonucleotide hybridization and fluorescent probes, melting temperature-based separation including single-stranded conformational polymorphism (SSCP), heteroduplex analysis, denaturing HPLC, and allele-specific amplification or primer extension to distinguish SNP alleles. TaqMan is also used using a two-dye reporter system and a reference dye. However, the cost of these probes can be considerable. This laboratory experiment will utilize allele specific amplification using one forward and two reverse allele-specific primers and SYBR Green I to detect the PCR products. The SNP alleles will be determined based on the melting temperature of the two possible amplicons.

Using the known DNA sequence of the human genome (or the genome of another organism of interest), including differences between alleles, an SNP can be probed using primers whose 3′ ends encompass the SNP. Allele-specific primers can be designed against any target containing a variation. This lab requires the use of three primers, two specific for each allele (reverse primers) and one that will function as the other primer that works with both allele-specific primers (forward primer) to amplify both alleles (Papp et al., 2003). The forward (5′) primer used should bind well to the complementary (3′) strands of both the allele variants.

The process used to detect the presence of the two alleles will be performed directly and simultaneously using real-time PCR and SYBR Green I dye. The fluorescent probe detects double-stranded DNA, and the real-time PCR instrument will be set to record the loss of fluorescence as the amplicon melts to a single-stranded DNA depending on the length and sequence of that DNA and thus the melting temperature of the probed sequences. Here, an 11-base pair GC-rich sequence (clamp) has been added to the 5′ end to extend the length and increase the melting temperature of one of the 3′ primers (Papp et al., 2003). The basis of the method is that oligonucleotides with a mismatched 3′ residue will not function as primers in PCR, even under appropriate conditions (Papp et al., 2003). Incorporation of a primer mismatch at the third base from the 3′ end of the primer has been shown to enhance the specificity of the PCR by further destabilizing the extension of the doubly mismatched primer. To additionally distinguish the allelic primers during amplification, different mismatches in the third 3′ base of both forward primers were employed. Alternatively, the two reverse primers could be labeled with different fluorescence dyes, although this approach was not taken. This adds significant expense to the PCR primers. The varying length method is a cost-effective method for the application and scoring of SNPs and introducing SNPs using a single fluorescent dye using 30 to 40 cycles of PCR. Optimal products are in the 50 to 150 bp range, and the amplicon produced in this experiment is 145 bp. The GC clamp has worked with a variety of sequences and lengths; a difference 4° C is discriminating. The expected melting temperature can be computed using the web tools at www.basic.northwestern.edu/biotools/oligocalc.html#helpbasic.

Although a number of SNP positions are available for analysis and are documented in the Short Tandem Repeat DNA Internet Database (STRbase), created and maintained by the National Institute of Standards and Technology (NIST), the chosen SNP for this purpose is mtDNA 16519 (Figure 14.1). In a study by Coble et al. (2004), the most highly variable SNP was found to be located at this position 16519 (T/C) between HV1 and HV2 on the mitochondrial DNA chromosome. Position 16519 varied in 9 of the 18 HV1/HV2 types.

In this experiment, the SNP base identity at mtDNA position 16519 will be evaluated for the C/T alleles for standard and unknown DNA using real-time PCR (Papp et al., 2003, Newton et al., 1989, Kleppe et al., 1971). The PCR primers that will be used in this experiment are shown in Table 14.1. Sample amplification and melt data are shown in Figures 14.2 and 14.3, respectively. The amplicons produced and shown here in the sample data were all

```
16381  TCAGATAGGG GTCCCTTGAC CACCATCCTC CGTGAAATCA ATATCCCGCA CAAGAGTGCT
       AGTCTATCCC CAGGGAACTG GTGGTAGGAG GCACTTTAGT TATAGGGCGT GTTCTCACGA

16441  ACTCTCCTCG CTCCGGGCCC ATAACACTTG GGGGTAGCTA AAGTGAACTG TATCCGACAT
       TGAGAGGAGC GAGGCCCGGG TATTGTGAAC CCCCATCGAT TTCACTTGAC ATAGGCTGTA

16501  CTGGTTCCTA CTTCAGGGTC ATAAAGCCTA AATAGCCCAC ACGTTCCCCT TAAATAAGAC
       GACCAAGGAT GAAGTCCCAG TATTTCGGAT TTATCGGGTG TGCAAGGGGA ATTTATTCTG
```

FIGURE 14.1   145 bp amplicon produced by PCR primers for mtDNA SNP 16519 (bold) (Vallone et al., 2004). Sequence is rCRS (NCBI Accession: NC_012920) (Andrews et al., 1999). Primers are underlined.

TABLE 14.1   PCR Primers for mtDNA SNP 16519

Annealing Location	Primer Sequence	Tm
5′ primer (NIST)	5′-ACCACCATCCTCCGTGAAAT-3′	61.6 °C
3′ primer 1	5′-CGTGTGGGCTATTTAGGCTTTAcGA-3′	58.9 °C
3′ primer 2 (with GC-handle)	5′- CGCGGCCGGCC- CGTGTGGGCTATTTAGGCTTTAcGG-3′	59.7 °C

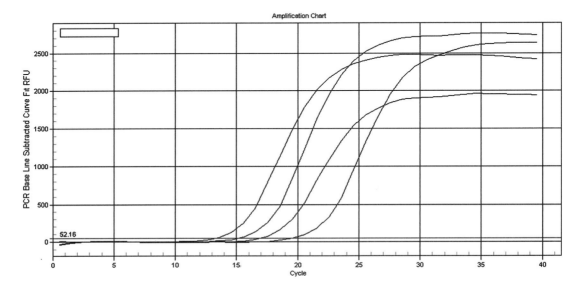

FIGURE 14.2   Exponential amplification curves that show the results of forty cycles of RT-PCR using the mtDNA SNP primers depicted in Table 14.1. Curves of standard DNA including K562, 9947A, and NIST SRM 2372, all at 1-ng template concentration, and an unknown as detected by SYBR Green I.

16519C. This is a 1-week laboratory with an optional second week to run an agarose gel to check for the production of the amplicon. A sample gel is shown in Figure 14.4.

PROCEDURE

Week 1

Real-Time PCR

1. Pipet the following into separate 1.5-mL microcentrifuge tubes using the primers for the mtDNA SNP 16519 T/C. (Student-designed primers for a SNP of interest or other published SNP primers may also be used if the 3′ end of one primer used to probe the SNP site is attached to the GC handle to provide T_m separation.)

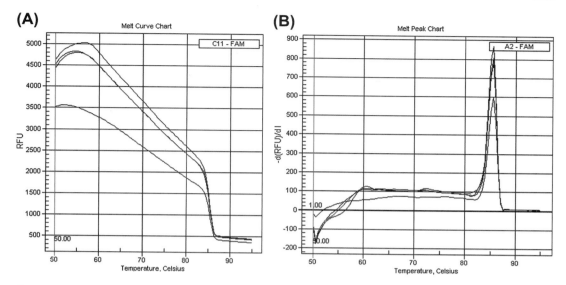

FIGURE 14.3 A. Melting curve plot from 50° to 95° C showing the temperature inflection at 85.5° C (left). B. First derivative plot of the melting curve from 50° to 95° C showing the predominant peak at 85.5° C for the K562, 9947A, NIST SRM 2372, and an unknown sample DNA (right). All standards used were 16519C, allowing the unknown to be assigned as 16519C.

12.5 μL of 2x iQ SYBR Green SuperMix
1 μL of 5 μM forward primer
1 μL of 5 μM reverse primer 1
1 μL of 5 μM reverse primer 2
8.5 μL of nuclease – free water
1 μL of commercial template DNA (1 to 2.5 ng so 1 μL of 1 ng/μL K562 DNA, 9947A, 9948, 007 DNA, NIST SRM 2372, simulated evidence DNA or other)25 μL Total Reaction Volume

2. Prepare at least one negative no template control sample (substitute nuclease free water for template DNA) (as a class for a single primer set or one per primer set).
3. Prepare simulated evidence samples, if desired, using student-extracted DNA to compare to standard DNA with known genotype.
4. Mix samples by pipetting up and down.
5. Pipette your samples into the 96-well plate by applying the liquid to the size of the well; carefully record the wells used. Pipette the sample into the bottom of the well, and do not release the pipette until withdrawn from the sample so that you will not produce bubbles.
6. When all student samples have been loaded, use the plastic sealing device to seal Microseal plastic covers over all samples.
7. Open the PCR machine for the instrument, and insert the well plate; fully close the lid.
8. Turn on the computer attached to the BioRad iQ5 RT-PCR instrument. Turn on the BioRad iQ5 RT-PCR instrument; wait for the self-test to complete (approximately 10 minutes prior to use).

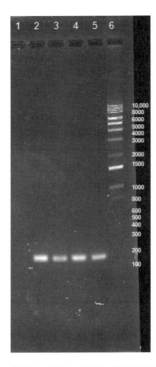

FIGURE 14.4   Sample 2% agarose gel of PCR of K562 DNA, 9947A, NIST SRM 2372, and unknown DNA using primers for mtDNA SNP 16519T/C. Lanes 2 to 5 contain the samples with the 156 bp amplicon. Lane 5 shows the DNA Logic Ladder (sizes indicated).

9. When the User Log-on screen appears, open the iQ5 v.2.0 software on the computer.
10. Under the Setup tab, select Protocol and select Create New to set up a new experiment. In the table at the bottom of the screen, edit the cycle settings to read Cycle 1, Step 1, 3:00 minute Dwell time, 95.0 Setpoint. After Cycle 1 is created, click on the … under Options on the Cycle 1 line to insert cycles "After," select "2 Steps," and select insert. Edit Cycle 2, Step 1 to read 0.15 Dwell time at 95.0 Setpoint and Cycle 2: Step 2 to read 1:00 Dwell time and 62.3 Setpoint (for 16519 primers) or select the gradient box to produce an eight-step gradient around a central temperature value (e.g., 50° to 65° C) (e.g., for student-designed primers). After Cycle 2 is created, click on the … under Options on the Cycle 2 line to insert cycles "After," select "1 Step," and select insert. In the drop-down box under PCR/Melt Data Acquisition, select Melt Curve. Edit the preset conditions to read 50.0 Setpoint. This will execute a melting curve of 91 cycles for 30 seconds each of melting every 0.5° C from 50° C. Select Save and Exit Protocol Editing; add a filename and click save.
11. The BioRad iQ5 automatically calculates the negative derivative of the change in fluorescence. When graphed, this yields a peak at the Tm of the PCR product. The Tm of the C variant produced by the reverse primer 2 will have the higher Tm due to the GC-rich handle.

Week 2

Agarose Gel Electrophoresis (Optional Week 2)

12. To prepare a 2% agarose gel, combine 1 g of agarose and 50 mL of 1x TAE buffer in an Erlenmeyer flask covered with plastic wrap punctured with a small hole.
13. Heat in the microwave for 40 seconds or until it boils and agarose is fully dissolved. Remove from the microwave.
14. Wait for it to cool to the touch, then pour it into the gel box positioned so that a rectangular box with four sides is created. Add a 14-well comb.
15. Once the gel is set (translucent blue color), rotate the gel so that the wells are positioned nearest the negative pole (black).
16. Fill the gel box with 1x TAE buffer to cover the gel.
17. Carefully, remove the comb from the gel.
18. To prepare samples for the gel, add 3 μL of the 6X gel loading buffer to each sample in the PCR plate.
19. Load 12 μL of the amplified samples-loading buffer to separate middle wells carefully by pipetting. Do not release the plunger until the pipette is removed from the gel so as to not draw up the samples.
20. To prepare the ladder for the gel, combine 8 μL of the DNA Logic ladder with 1 μL of 100x SYBR Green I dye and 1 μL of 6X loading dye, and combine 4 μL of the Lonza 20 bp ladder with 1 μL of 100x SYBR Green I dye and 1 μL of the 6X loading dye.
21. Load ladders to two lanes on opposite sides of the gel.
22. Place the safety cover on the gel box to run the sample toward the positive (red) electrode. Connect and turn on the power supply.
23. Run the gel for 30 to 60 minutes at no more than 15 V/cm (or 150 V for 10 cm gel).
24. Turn off the power supply and unplug it. Remove the gel safety cover.
25. While wearing gloves, remove the gel from the casting plate and place it on a UV light box, and take a picture using a digital photodocumentation system with a SYBR filter.
26. The C variant should not migrate as far as the T variant because of the GC handle on reverse primer 2.

QUESTIONS

1. What were the expected and observed T_m values for the C/T amplicons? Give reasons for any deviations.
2. From the gel, estimate the size of the amplicon produced by measuring the distances of the ladder fragments and the amplicon fragment(s), plotting log bp versus distance for the ladder, and using the line equation to compute the size of the amplicon.

References

Andrews, R.M., Kubacka, I., Chinnery, P.F., Lightowlers, R.N., Turnbull, D.M., Howell, N., 1999. Reanalysis and revision of the Cambridge reference sequence for human mitochondrial DNA. Nat. Genet. 23, 147.

Coble, M.D., Just, R.S., O'Callaghan, J.E., Letmanyi, I.H., Peterson, C.T., Parsons, T.J., 2004. Single nucleotide polymorphisms over the entire mtDNA genome that increase the forensic power of testing in Caucasians. Int. J. Legal Med. 118, 137–146.

Kleppe, K., Ohtsuka, E., Kleppe, R., Molineux, I., Khorana, H.G., 1971. XCVI. Repair replication of short synthetic DNA's as catalyzed by DNA polymerases. J. Mol. Biol. 56, 341–361.

Newton, C.R., Graham, A., Heptinstall, L.E., Powell, S.J., Summers, C., Kalsheker, N., Smith, J.C., Markham, A.F., 1989. Analysis of any point mutation in DNA. The amplification refractory mutation system (ARMS). Nucleic Acids Res. 17, 2503–2516.

Papp, A.C., Pinsonneault, J.K., Cooke, G., Sadée, W., 2003. Single nucleotide polymorphism genotyping using Allele-Specific PCR and fluorescence melting curves. BioTechniques 34, 1068–1072.

Vallone, P.M., Just, R.M., Coble, M.D., Butler, J.M., Parsons, T.J., 2004. A multiplex allele-specific primer extension assay for forensically informative SNPs distributed throughout the mitochondrial genome. Int. J. Legal Med. 118, 147–157.

Analysis of Deoxyribonucleic Acid (DNA) Sequence Data Using BioEdit

OBJECTIVE

To learn how to use free biological sequence editor (BioEdit) software for analyzing DNA sequence electropherograms.

SAFETY

No special precautions. This is an *in silico* lab.

MATERIALS

1. Computer with BioEdit freeware

BACKGROUND

Capillary electrophoresis (CE) data produced from genetic analyzers can be viewed with Applied Biosystems' Sequence Scanner software, biological sequence editor (BioEdit), Sequencher, and other programs. This experiment is designed to briefly introduce you experimentally collected DNA sequence data saved in .abi formats produced from the ABI 310 Genetic Analyzer or a similar capillary electrophoresis (CE) instrument. DNA sequencing is routinely performed when analyzing mitochondrial DNA samples using the highly variable (HV) 1, 2, and 3 regions. Mitochondrial DNA is often employed in familial analysis through maternal lines. DNA sequencing is a method of determining the exact DNA base or series of bases at a given location in the genome. The goal of this laboratory is to familiarize you with the analyst and computer roles of allele calls and sequence interpretation and possible artifacts including matrix effects, pull-up, major/minor peak analysis, mixture identification, heteroplasmy, too much or too little template, and degraded DNA.

Sanger et al. (1974) developed the process of DNA sequencing now referred to as Sanger dideoxy sequencing. DNA sequencing using the Sanger method employs a DNA template, a sequencing primer, DNA polymerase, $2'$-deoxy nucleotide triphosphates (dNTPs), $2'$, $3'$-dideoxynucleotides (nucleotide base analogs that lack the $3'$-hydroxyl group required for phosphodiester bond formation) (ddNTPs), and a reaction buffer. Whereas Sanger sequencing used radioactively labeled nucleotides, modern Big Dye chemistry from Applied Biosystems uses fluorescently labeled ddNTP bases. The fluorescent dyes each have separate excitation and emission wavelengths and can be discovered using a detector. In four separate reactions with ddA, ddC, ddG, or ddT, the labeled ddNTP base is randomly inserted into the sequence at the appropriate position to complementarily base pair with the template. When a dNTP is added to the $3'$ end, chain extension can continue. However, when a ddNTP is inserted, the chain extension halts because of the lack of the $3'$-OH in the sugar. Sequencing reactions result in the formation of extension products of various lengths terminated with ddNTPs at the $3'$ end. After removing the unincorporated ddNTPs and dNTPs from the reactions, the extension products are separated using CE using POP-6 polymer. The fragments are separated by size, and the smallest fragments elute first from the capillary. The sequence can be deduced based on the order of elution and detection of the fluorescent dyes on the labeled ddNTPs.

This lab will follow the SWGDAM reporting guidelines for DNA sequence data (found at www.fbi.gov/hq/lab/fsc/backissu/april2003/swgdammitodna.htm). If there are two or more nucleotide differences between the questioned and known samples, the samples can be excluded as originating from the same person or maternal lineage. If there is one nucleotide difference between the questioned and known samples, the result will be reported as inconclusive. Finally, if the sequences from questioned and known samples under comparison have a common base at each position or a common length variant in the HV2 C-stretch, the samples cannot be excluded as originating from the same person or maternal lineage.

PROCEDURE

BioEdit for DNA Sequence Analysis

1. Download BioEdit from www.mbio.ncsu.edu/bioedit/bioedit.html.
2. Double-click on the zipped file to extract and install.
3. The sample.abi file is downloaded with BioEdit. Obtain other sequence files of mtDNA sequence data from your instructor. Other sequence files may be downloaded from www.elimbio.com/sequencing_sample_files.htm and www.megasoftware.net/examples.php or www.dna-fingerprint.com/modules.php?op=modload&name=Downloads&file=index&req=viewdownload&cid=3.
4. Open BioEdit by clicking on the icon (Figure 15.1).

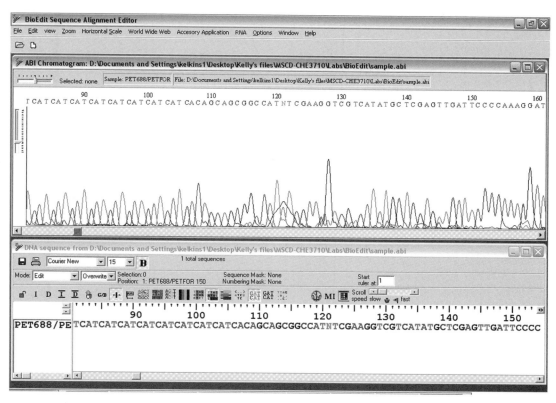

FIGURE 15.1 Biological Sequence Editor (BioEdit) User Interface.

5. Select File > Open > sample.abi. You will see a DNA sequence based on the automated nucleotide assignments and the raw electropherogram data provided by Applied Biosystems (ABI).

6. Note that some of the assignments are not A, T, G, or C; rather they are "N" because the software could not make the call given the operator or default parameters. You can assign these by highlighting the base in the sequence window and changing the base letter designation. (A is green, C is blue, T is red, and G is black.)

7. Make some observations about the sample sequence. Is it a mixture? Compare and contrast your ability to identify the base of interest if there is a stream of identical bases (poly-C stretch, for instance) versus as stretch in which the base varies. How long is the readable sequence (to what base pair number)? Is the middle, end or beginning of the sequence most reliably called? What are your observations about color overlap/deconvolution by the matrix? How might color deconvolution be improved? What is the source of the multiple peaks and big peaks over little peaks (e.g., what is the forensic term for this phenomenon)? What is the source of the peak height variation? What might be the cause if all peaks are barely above baseline? What might be the cause if all of the peaks are

outside the readable scale range of intensity units? Call the bases at positions 13, 121, and 720. How might analyst bias influence your selections? Identify the overall sequence in your report.

8. Repeat step 7 for the mitochondrial DNA sequences provided by your instructor. Assign the sequence of the provided HV I, II, and III regions. Report differences in the nucleotide sequence as compared to the revised Cambridge Reference Sequence (rCRS) (Andrews et al., 1999) using SWGDAM reporting guidelines.

QUESTIONS

Answer the questions in the procedure, and assign the sequence and differences as compared to rCRS for sequences provided.

References

BioEdit, www.mbio.ncsu.edu/BioEdit/bioedit.html, accessed 4-24-2012.

Andrews, R.M., Kubacka, I., Chinnery, P.F., Lightowlers, R.N., Turnbull, D.M., Howell, N., 1999. Reanalysis and revision of the Cambridge reference sequence for human mitochondrial DNA. Nat. Genet. 23, 147.

Sanger, F., Donelson, J.E., Coulson, A.R., Kösse, l.H., Fischer, D., 1974. Determination of a nucleotide sequence in bacteriophage f1 DNA by primed synthesis with DNA polymerase. J. Mol. Biol. 90, 315–333.

16

Ribonucleic Acid (RNA) Extraction

OBJECTIVE

To learn how to extract and analyze messenger RNA (mRNA) from simulated crime scene samples in order to differentiate body fluids.

SAFETY

Wear ethanol-treated gloves when handling any reagents so as to not contaminate them with nucleases. Avoid contact of skin with reagents.

MATERIALS

1. Dounce homogenizer
2. Mortar and pestle
3. 1-cc syringe with 18- to 21-gauge needles
4. Phosphate buffered saline (PBS) to wash cells
5. Extraction buffer (0.1 M NaCl, 10 mM Tris—HCl (pH 8), 40 mM dithiothreitol (DTT), 10 mM ethylenediamine tetraacetic acid (EDTA, pH 8), 70 mM sodium dodecyl sulfate (SDS), 0.65 mg/mL Proteinase K) prepared in nuclease-free water
6. Denaturation solution (4 M guanidine isothiocyanate, 0.02 M sodium citrate, 0.5% sarkosyl, 0.1 M 2-mercaptoethanol)
7. Spin-Ease extraction tubes
8. Phenol:Chloroform (5:1 solution, pH 4.5)
9. 2 M sodium acetate
10. 1X TE buffer (10 mM Tris, 0.1 mM EDTA, pH 8)
11. 1 M DTT
12. Nuclease-free water
13. Various variable volume micropipettes (e.g., 0.5 to 10 μL, 10 to 100 μL, 100 to 1000 μL)
14. Autoclaved pipette tips for pipettes
15. Sterile 1.5-mL microcentrifuge tubes
16. Gloves, lab coats, and goggles

17. Heating block or water bath
18. TURBO DNase (RNase-Free, 2 U/μL)
19. RNAsecure Resuspension Solution
20. 100% ethanol
21. 70% ethanol
22. GlycoBlue glycogen carrier
23. Isopropanol
24. 75% ethanol/25% diethylpyrocarbonate (DEPC)-treated water
25. RNA AWAY Reagent (Invitrogen)
26. PCR Primers (IDT)
27. OR (Procedure 2) Micro Fast-Track mRNA Isolation Kit (Invitrogen)
28. iScript One-Step RT-PCR Kit with SYBR Green, Random Decamer Primers, iScript reverse transcriptase enzyme mixture (iScript One-Step RT-PCR Kit With SYBR Green, BioRad, Instruction Manual)
29. Dry ice
30. Thermocycler
31. Agarose
32. Gel box and power supply
33. 100X SYBR and DNA ladder
34. 6X gel loading dye
35. Photodocumentation system with SYBR Green I filter

BACKGROUND

Body fluid stains recovered from crime scenes may be homogenous (e.g., blood from one individual) or heterogeneous (e.g. blood from multiple individuals or blood and semen from the same or different individuals) in nature. A sample may contain blood from multiple individuals or blood and semen from the same or different individuals. Although differential extraction used to remove DNA from semen and DNA from another source (e.g., vaginal cells) in the same sample has played an important role in DNA typing for a long time, investigators may ask lab analysts to determine the tissue source of biological material in an investigation. Serological and presumptive tests have long played a role in these determinations but are time and material consumptive. Attenuated total reflectance Fourier transform infrared spectroscopy and Raman spectroscopy have been proposed for onsite, rapid body fluids differentiation and detection but are not locus specific as is RNA or DNA typing.

RNA analysis also plays a role in tissue differentiation. Different cell types produce different messenger RNAs (mRNAs) depending on environment, stress, and other activity levels. This method can be used to differentiate semen, blood, saliva, and vaginal secretions present in a sample. This information can be used to determine ratios of body fluids present in a sample.

Mammalian cells contain approximately 10 μg of RNA, and of that, 1% to 5% is polyA+ mRNA. The rest of the RNA is ribosomal RNA (rRNA) (80% to 85%) and transfer RNA (tRNA) (15% to 20%). After gently lysing the cells and inactivating RNases, the RNA can be separated from the DNA using affinity chromatography with oligo(dT) cellulose columns

TABLE 16.1 PCR Primers for Tissue and Fluids Differentiation

Tissue	Locus	Forward primer	Reverse Primer
Blood	beta-spectrin	5'-AGG-ATGGCT-TGG-CCT-TTA-AT-3'	5'-ACT-GCCAGC-ACC-TTC-ATC-TT-3'
Saliva	histatin 3	5'-GCA-AAG-AGA-CAT-CAT-GGG-TA-3'	5'-GCC-AGT-CAA-ACC-TCC-ATA-ATC-3'
Semen	protamine 1	5'-GTC-CGA-TAC-CGCGTG-AGG-AGC-CTG-3'	5'-GCC-TTC-TGCATG-TTC-TCT-TCC-TGG-3'
Vaginal secretions	mucin 4	5'-GGA-CCA-CAT-TTT-ATCAGG-AA-3'	5'-TAG-AGA-AAC-AGG-GCATAG-GA-3'

to bind the polyA+ RNA under high-salt conditions. The rRNA is removed using a low-salt buffer, and the mRNA is eluted using a very low ionic strength buffer.

The RNA will be extracted using either a standard guanidine isothiocyanate-phenol-chloroform extraction (Ausubel et al., 1990, Sambrook et al., 1989) or a Micro Fast-Track mRNA Isolation kit (Invitrogen) in this laboratory. Because of the limited sample size, DNA and mRNA are co-extracted from forensic samples for this procedure (Alvarez et al., 2004).

Using the extracted mRNA, cDNA will be synthesized by PCR using random decamer primers. A subsequent PCR will be performed with the cDNA using PCR primers for tissue and body fluids differentiation (Table 16.1) and an agarose gel will be used to separate the amplicons to differentiate blood, semen, saliva, and vaginal secretions.

PROCEDURE

When working with RNA, sterile and RNase-free conditions are absolutely essential to recovering RNA from your product. Only disposable, individually wrapped, RNase-free, sterile plastic ware should be used. Only new, sterile pipette tips (filter tips preferred) and microcentrifuge tubes should be used. Any glassware used for working with RNA should be cleaned with detergent, thoroughly rinsed, autoclaved, and oven baked at >210° C for at least 3 hours prior to use. Gloves should always be used to handle reagents and RNA samples to prevent contamination from RNases on the surface of the skin.

Option 1

Week 1 (Gd:Phenol:Chloroform Extraction)

1. Obtain a simulated crime scene sample or swab from the instructor. Clean workspace with RNase AWAY Reagent to remove RNase contamination from surfaces.
2. Incubate the sample at 56° C for 1 hour in 500 μL of extraction buffer. (Add an additional 0.39 M DTT to the extraction buffer for semen stains.)
3. Remove the sample from the extraction buffer.
4. Place the sample in a Spin-Ease extraction tube and centrifuge at 16,000 xg for 5 minutes.
5. Discard the filter and fabric or swab remnants, if any.

6. Add 50 μL 2 M sodium acetate and 600 μL 5:1 Phenol:Chloroform acid solution, vortex vigorously, and incubate at 4° C at least 20 minutes or until phases separate and layers are visible.
7. Centrifuge at 16,000 x g for 20 minutes. The aqueous phase contains the DNA and RNA.
8. Transfer 200 μL of the aqueous phase to a sterile 1.5-mL microcentrifuge tube labeled "DNA" and 200 μL of the aqueous phase to a sterile 1.5-mL microcentrifuge tube labeled "RNA."
   (Perform the steps for DNA extraction, if desired.)
9. To precipitate the DNA, add 500 μL ice-cold 100% ethanol to the "DNA" tube and let stand at −20° C for 1 hour or overnight (per your instructor). Do not discard.
10. To isolate the RNA, add 2 μL of GlycoBlue glycogen carrier to the "RNA" tube.
11. Add 250 μL of isopropanol to the "RNA" tube to precipitate the RNA, and let stand at −20° C for 1 hour or overnight. Do not discard.

Week 2 (Gd:Phenol:Chloroform Extraction)

12. After the DNA incubation, centrifuge for 15 minutes at 16,000 xg.
13. Carefully remove the supernatant.
14. Wash the (invisible) DNA pellet with 1 mL of 70% ethanol (room temperature) and then centrifuge for 5 minutes at 16,000 xg.
15. Vacuum dry the DNA product for 4 to 7 minutes, and resolubilize the pellet in 50 μL TE buffer.
16. Incubate the DNA samples for 45 minutes at 56° C. Store DNA samples at 4° C.
17. Centrifuge the RNA samples at 16,000 xg for 20 minutes. Carefully pipette off the supernatant and discard.
18. Wash the RNA pellet with 1 mL of 75% ethanol/25% diethylpyrocarbonate (DEPC)-treated water, and then centrifuge for 10 minutes at 16,000 xg. Remove the supernatant by carefully pipetting and discard.
19. Vacuum-dry the RNA product for 3 to 5 minutes, and resolubilize the pellet in 20 μL RNAsecure Resuspension Solution.
20. Incubate the RNA samples for 10 minutes at 60° C.
21. Treat the RNA sample with 3 μL (6 U) of TURBO DNase (RNase-Free, 2 U/μL) and incubate at 37° C for 2 to 3 hours.
22. Inactivate the DNase by incubating the RNA samples at 75° C for 10 minutes, then chill the samples on ice. Store the RNA samples at −20° C.
23. The samples can be quantified using the RiboGreen method for RNA and PicoGreen, UV-Vis, fluorescence, real-time PCR, TaqMan, or QuantiBlot for DNA, as desired.

Option 2

Week 1 (Micro-Fast Track 2.0 Kit)

1. Obtain a simulated crime scene sample or swab from the instructor. Clean the workspace with RNase AWAY Reagent to remove RNase contamination from needed surfaces.
2. Immediately before use, prepare the Lysis Buffer. If there is a white precipitate in the SDS in the Stock Buffer, heat the solution to 65° C until the precipitate is dissolved. Cool the

solution to room temperature. Add 2 µL of Proteinase K (20 mg/mL) used to degrade proteins and RNases to 1 mL of Stock Buffer for each sample. (Multiply by the number of class samples to prepare this reagent in bulk. Any prepared reagent must be used this session.)

3. Wash the swab/fabric cell sample in 4° C phosphate buffered saline (PBS) solution in a 15-mL sterile, conical centrifuge tube. Remove the fabric or swab, if any.

4. Centrifuge the cells. Resuspend the pellet in 1 mL of PBS.

5. Transfer the cells to a sterile microcentrifuge tube and centrifuge the cells to pellet. Pipet off the PBS carefully.

6. Add 1 mL of Micro-FastTrack 2 Lysis Buffer containing Proteinase K to the pellet to lyse the cells. Vortex until cells are completely resuspended.

7. Use a sterile, plastic, 1 cubic centimeter (cc) syringe fitted with a 18- to 21-gauge needle to shear the DNA by drawing the cell lysate into the syringe two to four times. Continue until the solution is less viscous (up to 10 times).

8. Incubate the sample at 45° C for 20 minutes. Centrifuge at 4000 xg at room temperature for 5 minutes to remove the insoluble material. Transfer the supernatant to a new, sterile, microcentrifuge tube.

9. Add 63 µL of 5 M NaCl stock solution to each 1 mL of sample containing 0.2 M NaCl to bring the NaCl concentration up to 0.5 M. Mix well.

10. Use a sterile, plastic, 1-cc syringe fitted with a 18- to 21-gauge needle to shear any remaining DNA by drawing the cell lysate into the syringe two to four times. Continue until the solution is less viscous (up to 10 times). Shearing the DNA will yield a cleaner RNA preparation.

11. Obtain a vial of oligo(dT) cellulose (from the desiccator) and add your sample to the vial and label clearly. Cover to seal. Allow the cellulose to swell for 2 minutes.

12. Using a rocking platform, rock the tube at room temperature for 15 to 60 minutes (longer times will increase mRNA yield).

13. Centrifuge the sample-oligo(dT) cellulose mixture for 5 minutes at 4000 xg at room temperature.

14. Carefully remove the supernatant carefully from the resin bed. Do not disturb the oligo(dT) cellulose pellet.

15. Resuspend the oligo(dT) cellulose in 1.3 mL of Binding Buffer. Centrifuge at 4000 xg in a microcentrifuge for 5 minutes at room temperature. Remove the Binding Buffer from the oligo(dT) cellulose resin bed. Repeat this step twice more.

16. Add 0.3 mL of Binding Buffer to the oligo(dT) cellulose resin, and transfer some of the sample (until it is full) to a spin column. (These come as a spin column/microcentrifuge tube set.)

17. Centrifuge for 10 seconds at 4000 xg at room temperature. Repeat until all sample and resin has been loaded onto the spin column. Discard the eluent.

18. Add 500 µL of Binding Buffer to the spin column on the microcentrifuge tube.

19. Centrifuge for 10 seconds at 4000 × g at room temperature. Continue to wash the sample and resin three more times with Binding Buffer. Use a UV-Vis instrument to read the absorbance at 260 nm of the flow-through. Keep washing the sample resin with Binding Buffer until the A260 nm is less than 0.05.

20. Add 200 µL of Low Salt Wash Buffer to remove SDS and nonadenylated RNAs such as rRNAs. Gently mix the sample resin to resuspend with a sterile pipette tip. Be very

careful not to damage the membrane of the spin column as you perform this step. This will cause you to lose your sample.

21. Centrifuge for 10 seconds at 4000 xg at room temperature. Wash the sample once more with Low Salt Wash Buffer and centrifuge again for 10 seconds at 4000 xg.

22. Transfer the spin column into a new, sterile, RNase-free microcentrifuge tube (provided in the kit).

23. Add 100 µL of Elution Buffer. Gently mix the sample resin to resuspend with a sterile pipette tip. Be very careful not to damage the membrane of the spin column as you perform this step. This will cause you to lose your sample.

24. Centrifuge for 10 seconds at 4000 xg at room temperature to elute your mRNA into the new tube. *Do not throw the eluent liquid away. Repeat steps 23 to 24, and keep the liquid eluent.*

25. Remove the spin column from the microcentrifuge tube. The tube now contains ~200 µL of RNA. This is your mRNA sample. You can continue to spin the spin column to improve your yield if you do not have 200 µL.

26. Add 10 µL of 2 mg/mL glycogen carrier protein (supplied in kit), 30 µL of 2 M sodium acetate (supplied in kit), and 600 µL of 100% (200 proof) ethanol to precipitate the mRNA. Place the sample(s) on dry ice for 10 to 15 minutes.

27. Centrifuge for 15 minutes at 4° C at maximum speed in a microcentrifuge. Carefully remove the supernatant without disturbing the pellet. Discard the supernatant. *The mRNA is in the pellet.* Do not discard the pellet.

28. Add 70% ethanol to wash the pellet. Centrifuge and remove the ethanol supernatant.

29. Allow the pellet to air-dry with the microcentrifuge cap open for 5 to 10 minutes.

30. Add 1 to 10 µL of Elution Buffer (10 mM Tris-HCl, pH 7.5) to resuspend the mRNA. Use the mRNA immediately or store at −80° C.

Week 2 (cDNA Synthesis by PCR)

1. Pipet 80 ng (or, if less, the entire extraction) of mRNA from your yield. Heat the mRNA sample for 3 minutes at 75 ° C.

2. To a thin-walled PCR tube or 96-well plate for each sample, add the following:
   12.5 µL iScript One-Step RT-PCR Kit with SYBR Green (BioRad) reaction mix
   1 µL iScript reverse transcriptase enzyme mixture
   2 µL of 50 µM Random Decamer Primers (Ambion)
   8.5 µL of nuclease-free water
   1 µL RNA extract (1 pg to 100 ng)
   25 µL Total Reaction Volume (add mineral oil if thermocycler does not have a heated lid)

3. Prepare one negative (no template) control sample (substitute nuclease-free water for template DNA) (as a class).

4. Mix all tubes thoroughly by inverting or flicking tubes.

5. Centrifuge to remove bubbles. (If using 96-well plates, mixing may be performed using a sterile pipette tip but care should be used to not introduce bubbles unless a plate centrifuge is available.)

6. Place all the tubes in the PCR Thermal Cycler, and begin the following preprogrammed cycles:
   50° C for 10 minutes

95° C for 5 minutes
 95° C for 10 seconds
 60° C for 30 seconds
Repeat for 30 to 40 cycles
95° C for 60 seconds
50° C for 60 seconds
50° to 95° C for 10 for seconds each in 0.5° C steps, 91 cycles

Hold at 4° C until the next morning. Your instructor will remove your samples and store them in the freezer for you until the next lab.

Week 3 (PCR)

1. Recover the amplified cDNA from the last lab.
2. To a thin-walled PCR tube or 96-well plate for each sample, add the following:
   12.5µL 2x iQ SYBR Green SuperMix (with Mg^{2+} and dNTPs)
   4 µL of 5 µM each Forward Primers Mix (Table 16.1)
   4 µL of 5 µM each Reverse Primers Mix (Table 16.1)
   3.5µL of nuclease-free water
   1 µL cDNA (2 to 6 ng for blood, semen, saliva, and 66 ng for vaginal secretions)
   25 µL Total Reaction Volume (add mineral oil if the thermocycler does not have a heated lid)
3. Prepare one negative (no template) control sample (substitute nuclease-free water for template DNA) (as a class).
4. Prepare one positive control sample (1 ng in 1 µL for template DNA of each body fluid type) (as a class).
5. Mix all tubes thoroughly by inverting or flicking tubes.
6. Centrifuge to remove bubbles. (If using 96-well plates, mixing may be performed using a sterile pipette tip, but care should be used to not introduce bubbles unless a plate centrifuge is available.)
7. Place all the tubes in the PCR Thermal Cycler, and begin the following preprogrammed cycles:
   95° C for 3 minutes
   95° C for 15 seconds
   55° C for 30 seconds
   72° C for 40 seconds (60 seconds per kb)
   Repeat the previous three-step cycle 30 to 40 times

Hold at 4° C until the next morning. Your instructor will remove your samples and store them in the freezer for you until the next lab.

At this point, the cDNA can also be used in a STR Typing kit PCR reaction.

Week 4 (Agarose Gel Electrophoresis)

1. Prepare a 2.5% agarose (MetaPhor or NuSieve GTG) gel. If preparing gel, combine 1.25 g of agarose and 50 mL of 1x TAE buffer (which offers better resolution of DNA fragments

than TBE buffer) in an Erlenmeyer flask covered with plastic wrap that has been punctured with a small hole.

2. Heat in the microwave for 40 seconds or until it boils and agarose is fully dissolved. Replace any loss of liquid with deionized water.

3. Remove from the microwave.

4. Wait for it to cool to the touch, then pour it into the gel box positioned so that a rectangular box with four sides is created. Add a 14-well comb.

5. Once the gel is set (translucent blue color), rotate the gel so that the wells are positioned nearest the negative pole (black).

6. Fill gel box with 1x TAE buffer to cover the gel.

7. Remove the comb from the gel.

8. To prepare samples for the gel, add 3 µL of the 6x loading dye to each 25 µL sample in the PCR tubes or 96-well plate.

9. Load 12 µL of the amplified samples-loading buffer to separate wells carefully by pipetting. Do not release the plunger until the pipette is removed from the gel so as to not draw up the samples.

10. To prepare a ladder for the gel, combine 8 µL of the DNA Logic ladder (or desired ladder) with 1 µL of 100x SYBR Green I dye and 1 µL of 6X loading dye, and combine 4 µL of the Lonza 20 bp ladder with 1 µL of 100x SYBR Green I dye and 1 µL of the 6X loading dye. Load the entire quantity of each ladder to two separate wells on opposite sides of the gel.

11. Place the safety cover on the gel box to run the sample toward the positive (red) electrode. Connect and turn on the power supply.

12. Run the gel for 90 to 180 minutes at no more than 15 V/cm (or 150 V for 10 cm gel).

13. Turn off the power supply and unplug it. Remove the gel safety cover.

14. While wearing gloves, remove the gel from the casting plate, place it on a UV light box, and take a picture using a digital photodocumentation system with a SYBR filter.

QUESTION

Identify the body fluid source using the gel data.

References

Alvarez, M., Juusola, J., Ballantyne, J., 2004. An mRNA and DNA co-isolation method for forensic casework samples. Analytical Biochemistry 335, 289–298.

Ausubel, F.M., Brent, R., Kingston, R.E., Moore, D.D., Seidman, J.G., Smith, J.A., Struhl, K. (Eds.), 1990. Current Protocols in Molecular Biology. Greene Publishing Associates and Wiley-Interscience, New York.

iScript™ One-Step RT-PCR Kit With SYBR® Green, BioRad, Instruction Manual.

Sambrook, J., Fritsch, E.F., Maniatis, T., 1989. Molecular Cloning: A Laboratory Manual, 2nd ed. Cold Spring Harbor Laboratory.

Y-STR Polymerase Chain Reaction (PCR) Deoxyribonucleic acid (DNA) Amplification and Typing

OBJECTIVE

To learn how to set up and run Y-STR PCR reactions for DNA amplification and fragment analysis by capillary electrophoresis and interpret the results.

SAFETY

Handle all reagents with ethanol-rinsed gloves to avoid contaminating the kit components. The PCR-primers are labeled with fluorescent dyes; avoid contact with your hands. Caution: High voltage is used in the capillary electrophoresis.

MATERIALS

1. Extracted DNA simulated evidence template (1- to 2.5-ng range)
2. AmpFlSTR YFiler PCR Amplification Kit complete with AmpliTaq Gold DNA Polymerase, PCR Reaction mix, and Primer set (Applied Biosystems)
3. Standard DNA (e.g., NIST SRM 2395, 007 DNA, 9948)
4. PCR thermal cycler
5. Micropipettes and sterile tips (filter tips preferred) (e.g., 0.5 to 10 μL, 2 to 20 μL, 10 to 100 μL, 100 to 1000 μL)
6. Nuclease-free water
7. Microcentrifuges
8. Microcentrifuge tube rack

9. Thin-walled PCR tube rack
10. Thin-walled PCR tubes
11. Disposable gloves
12. Thin line permanent marker
13. Vortexer

BACKGROUND

Sexual assault crimes constitute a large quantity of crimes for which DNA typing is pursued. Crime statistics estimate that males commit 80% of violent crimes and approximately 95% of sexual offenses. Sexual assault crimes account for more than 50% of biological evidence submitted and processed in casework laboratories as part of investigations for sexual assault.

The evidence that results and that is recovered typically consists of a mixture of male and female body fluids, namely semen and vaginal cells/fluid. Typically the male fraction is in a low quantity, but differential extraction has proved able to isolate the sperm cells from other cells. However, difficulties remain when profiling the azoospermic, oligospermic, or vasectomized male. Low levels of male DNA may also be isolated in evidence associated with violent crimes or weapon handles or in fingernail scrapings.

Y-chromosome short tandem repeat (Y-STR) markers are polymorphic DNA loci that contain a repeated nucleotide sequence located on the Y-chromosome. As the Y-chromosome is passed genetically from paternal grandfather to father to son, Y-STRs provide a means of paternal familial typing in addition to forensic missing persons and missing persons casework. Typically, tetranucleotide DNA repeats (e.g., TAGA repeats for DYS19) are probed. As there is a single copy of the Y-chromosome, the STR loci probed generally will yield a single allele or a haplotype. However, some of the loci will have more than one allele call because of gene duplication events that allow the primers to bind twice in two different regions (e.g., DYS385 a/b) or a single region in which the forward primers bind twice (e.g., DYS 389 I/II).

The AmpFlSTR YFiler PCR Amplification Kit is a Y-STR multiplex assay that employs a single PCR reaction to amplify 17 Y-STR loci (Mulero et al., 2006). The kit amplifies the loci found in Table 17.1. The alleles have been named in accordance with the recommendations of the DNA Commission of the International Society for Forensic Haemogenetics (ISFH). The recommended Y-STRs to be included for DNA typing by international organizations are listed in Table 17.2.

The 17 loci are separated by allele size, dye-labeled primer, or both, as indicated in Table 17.1. The fifth dye, LIZ (orange), is used to label the GeneScan-500 Size Standard. The size standard is designed for sizing DNA fragments in the 35 to 500 bp range, and it contains 16 single-stranded fragments of 35, 50, 75, 100, 139, 150, 160, 200, 250, 300, 340, 350, 400, 450, 490, and 500 bases. Each of the dyes has a unique excitation and emission maxima and emits fluorescence at a unique wavelength. Multicomponent analysis and diffraction gratings are used to differentiate the dyes even if the colors exhibit spectral overlap.

TABLE 17.1 Y-STR Loci Amplified by AmpFISTR® YFiler Kit and Allele Calls for DNA 007 and DNA 9948 Standards (Mulero et al., 2006, Applied Biosystems AmpFISTR YFiler PCR Amplification Kit User's Manual, DNA-fingerprint.com)

Locus	Alleles Included in Yfiler Kit Allelic Ladder	Dye Label in YFiler	DNA 007 Genotype	DNA 9948 Genotype
DYS456	13-18	6-FAM (blue)	15	17
DYS389I	10-15		13	13
DYS390	18-27		24	24
DYS389II	24-34		29	31
DYS458	14-20	VIC® (green)	17	18
DYS19	10-19		15	14
DYS385 a/b	7-25		11,14	11,14
DYS393	8-16	NED® (yellow)	13	13
DYS391	7-13		11	10
DYS439	8-15		12	12
DYS635	20-26		24	n/a
DYS392	7-18		13	13
Y GATA H4	8-13	PET® (red)	23	12
DYS437	13-17		15	15
DYS438	8-13		12	11
DYS448	17-24		19	19

TABLE 17.2 Y-STR Recommendations for DNA Typing

Recommended Y-STRs	Loci
European minimal haplotype	DYS19, DYS385a/b, DYS389I/II, DYS390, DYS391, DYS392, DYS393
Scientific Working Group-DNA Analysis Methods (SWGDAM)	European minimal haplotype plus DYS438 and DYS439
Additional highly polymorphic loci	DYS437, DYS448, DYS456, DYS458, DYS635 (Y GATA C4), and Y GATA H4

For the PCR reaction, the kit reagents include the YFiler primer sets for each of the loci, AmpliTaq Gold DNA polymerase (5 U/μL), and the reaction mix containing nucleotide triphosphates (dNTPs) including dATP, dTTP, dGTP, and dCTP, 0.05% sodium azide, $MgCl_2$, and bovine serum albumin, *Taq* DNA polymerase, and Tris buffer. For a positive control, DNA 007 or SRM 2395 male DNA samples are used. A negative control (female) DNA sample 9947A and the allele ladder for sizing are also included in the kit. The extracted DNA from a simulated evidence sample is the unknown or questioned sample. The kit works under a narrow range of conditions to be time- and cost-effective. A full profile is routinely produced using 1–2.5 ng of DNA after 30 cycles. All reagents should be kept on ice while preparing the PCR reaction, and PCR should be started immediately after adding all reagents to avoid mis-priming. The dye-labeled reagents should not be exposed to light for extended periods of time. All reagents should be stored in the freezer. If the thermal cycler does not have a heated top, the samples and controls should be topped with mineral oil to prevent evaporation. The thermal cycler with a variable heating block will cycle through the temperatures required for the reaction. The YFiler kit utilizes a three-step cycle (the cycle temperatures and times may vary for other kits). First, the AmpliTaq Gold is heat activated at 95° C for 11 minutes. Then the following cycle is carried out 30 times. (1) The DNA is denatured at 94° C for 1 minute. (2) The primers are allowed to anneal (base pair) at 61° C for 1 minute. (3) The Ampli*Taq* Gold polymerase attaches where the primer and DNA are bound and extends the primer based on the template DNA sequence at 72° C for 1 minute. A final extension of 45 minutes is run at 60° C to finish the elongation of all strands. The thermocycler will be set to refrigerate the samples (4° C) until we are able to return to them for freezing before separation using capillary electrophoresis. As an alternate to the YFiler or other commercial kit, the Y-STRs of interest can be amplified using a PCR reaction specific for them using published or designed and purchased primers and reaction mix.

The application of PCR-based typing for forensic or paternity casework requires validation studies and quality control measures prior to adoption by the crime lab. The quality of the purified DNA sample, as well as small changes in buffers, ionic strength, primer concentrations, choice of thermal cycler, and thermal cycling conditions, can affect the success of PCR amplification. Y-STR analysis is subject to contamination by very small amounts of nontemplate human DNA. Extreme care should be taken to avoid cross-contamination in preparing sample DNA, handling primer pairs, setting up amplification reactions, and analyzing amplification products. Reagents and materials used prior to amplification (e.g., STR 10X Buffer, 007 control DNA, and dye-labeled 10X primer pairs) should be stored separately from those used following amplification (e.g., dye-labeled allelic ladders, loading solutions, and Gel Tracking Dye). A negative no-template control reaction should always be included in a reaction set. It will act as a useful indicator of reagent contamination if it amplifies. Some of the reagents used in the analysis of STR products, including polymer and dye-labeled primers, are potentially hazardous and should be handled accordingly.

The YFiler allelic ladder contains the most common alleles for each locus. Genotypes are assigned by comparing the sizes obtained for the unknown samples with the sizes obtained for the alleles in the allelic ladder. The company has sequenced alleles contained in the allelic ladder and from population samples to verify both the length and repeat structure. (Routine

Y-STR analysis of unknown samples does not include sequencing.) The size range is the actual base pair size of the sequenced alleles including the 3' A+ nucleotide addition. The YFiler PCR Amplification Kit is designed so that a majority of the PCR products will contain the nontemplated 3' A+ nucleotide post-PCR.

In this lab assignment, we will amplify the Y-STR fragments at the YFiler loci to prepare samples to be subsequently separated using capillary electrophoresis and POP-4 polymer in another laboratory period (Chapter 12).

PROCEDURE

Week 1

Part A: Preparation of the DNA Template for PCR

Dilute previously extracted genomic DNA from known reference or simulated crime scene samples to 1 to 2.5 ng in 10 µL.

Part B: Preparation of the Positive Control Template (Known Template — 007 or Other)

The control reaction is prepared by preparing DNA in the same manner as that described in Part A. The only difference is that 10 µL of SRM 2395 or 007 or other standard DNA template (0.10 ng/µL) is added to the tube instead of the previously isolated genomic DNA.

Part C: Preparation of the Negative Control

The negative control is prepared by substituting 10 µL of sterile distilled, deionized water to the reaction tube so as to maintain the same volume as the previous two reaction tubes.

Part D: Preparation of the PCR Mixture Using a Commercial Multiplex Kit

1. Vortex the AmpF*l* STR PCR Reaction Mix, YFiler Primer Set, and AmpliTaq Gold DNA Polymerase for 5 seconds. Spin the tubes briefly in a microcentrifuge to remove any liquid from the caps.
2. Plan to prepare one negative control sample (substitute nuclease-free water for template DNA) (individually or as a class) and one positive control sample (substitute 007 DNA or SRM 2395, 0.10 ng/µL for template DNA) (individually or as a class).
3. Compute the number of total reactions to run, and prepare master mix for the number of reactions needed plus one by pipetting the following reagents into a 96-well plate or thin-walled microcentrifuge tubes:
   9.2 µL AmpFlSTR YFiler Kit PCR Reaction Mix
   5 µL AmpFlSTR YFiler Kit Primer Set
   0.8 µL AmpliTaq Gold DNA Polymerase
   10 µL Control DNA (0.1 ng/µL) or 10 µL of extracted template DNA (0.5 to 1.0 ng)
   25 µL Total Reaction Volume

4. Mix thoroughly by vortexing at medium speed for 5 seconds.
5. Spin the tube briefly in a microcentrifuge to remove any liquid from the cap.
6. If the thermocycler does not have a heated cap, add one drop of mineral oil.

Part E: The Temperature Cycle Used for YFiler PCR Amplification of the 17 STR Loci

Place all the tubes in the PCR Thermal Cycler, and program the following cycles:

1. 95° C for 11 minutes (initial incubation)
2. 94° C for 60 seconds (denature)
   61° C for 60 seconds (anneal)
   72° C for 60 seconds (extend)
   Repeat 30 times.
3. 60° C for 80 minutes (final extension to add 3' A+ nucleotide)
4. Hold at 4° C until the next morning.

Begin the PCR on the thermal cycler. Your instructor will remove the samples and store them in the freezer until next week.

Week 2 (or Performed by Instructor)

Refer to Chapter 12 for instructions on how to perform capillary electrophoresis post-PCR.

QUESTION

Report the Y-STR profile post-capillary electrophoresis for the questioned (unknown) and known samples.

References

Applied Biosystems AmpFlSTR® YFiler™ PCR Amplification Kit User's Manual, Applied Biosystems AmpFlSTR® YFiler™ PCR Amplification Kit User's Manual, www3.appliedbiosystems.com/cms/groups/applied_markets_support/documents/generaldocuments/cms_041477.pdf, (accessed 11-22-11.).

007 Genotype, DNA Fingerprint: www.dna-fingerprint.com/index.php?module=pagesetter&;tid=1&orderby=locus, allele&filter=template:like:007

9948 Genotype, DNA Fingerprint: www.dna-fingerprint.com/index.php?module=pagesetter&;tid=1&orderby=locus, allele&filter=template:like:9948

Mulero, J.J., Chang, C.W., Calandro, L.M., Green, R.L., Li, Y., Johnson, C.L., Hennessy, L.K., 2006. Development and validation of the AmpFlSTR Yfiler PCR amplification kit: a male specific, single amplification 17 Y-STR multiplex system. J. Forensic Sci. 51, 64–75.

Human Genetic Analysis: Paternity or Missing Persons Cases and Statistics

OBJECTIVE

To learn how to interpret DNA typing data in paternity and missing persons cases using statistics.

SAFETY

No special safety precautions. This is an *in silico* lab.

MATERIALS

1. Computer with Microsoft Excel (or equivalent) or calculator

BACKGROUND

Paternity cases and missing persons cases can be criminal and domestic in nature. They differ from other forensic cases in that these individuals are most often related. For example, the most common paternity case includes one parent and a child who are evaluated for the purpose of assessing an alleged parent. Reference samples must be collected and submitted from the child, known parent, and alleged parent. In other, rarer cases, it may be necessary to evaluate by genetic testing that two individuals are the biological parents of a person who is the source of a stain, swab, or other evidence (criminal parentage). Paternity testing may also be used in mass disaster cases in which a paternal member of the family unit is missing or in immigration cases. Reference samples will be collected and analyzed from the available male, biologically related, family members with the goal of including or excluding an evidential sample with a possibility of being included in the family unit. If the case involves a missing child, both biological parents will be asked to submit DNA samples. In rare cases such as cases of abandonment, it may be necessary to evaluate a child and an alleged mother. In this case, reference samples must be submitted from the child and the alleged mother (and biological father, if known). Samples from the maternal and paternal grandparents may also be collected and analyzed if samples cannot be obtained from the biological mother or father.

In this experiment, you will determine the partial DNA profile for a "missing person" from the analysis of close family members. DNA analysts often have to recreate genotypes for those whose DNA is not readily available for analysis. A recent case of great national interest involved the identification of the remains of the Vietnam soldier who had been interred in the Tomb of the Unknown Soldier.

Biological offspring of two parents inherit half of their autosomal chromosomes and autosomal alleles from each parent. They also inherit one sex chromosome from each parent. The two haploid sets from gametes combine to form a diploid set in autosomal cells. In terms of paternity, the child, a son or a daughter, will inherit half of the father's autosomal chromosomes and the son will inherit a Y-chromosome from the father, whereas the daughter will inherit the father's X-chromosome. Allele profiles from the haploid X- and Y-chromosome markers are especially useful in these cases. The male offspring will match the biological father at all Y-chromosome loci, whereas the female offspring will match the biological father at all X-chromosome loci (Table 18.1). Typically the X and Y markers are combined with autosomal markers. Statistics can help to determine if an alleged father should be included or excluded as the father of the offspring. Because no known phenotypes are associated with STR loci, you need only consider the genotypes of the parents and their offspring.

The statistical calculation is used in forensic biology to assess the strength of evidence and the probability that there could be a random match to some other person. In paternity cases, likelihood ratios and combined probability of exclusion are two calculations that can be performed to evaluate the certainty of the evidence. To determine the likelihood that a specific genotype will occur in a population, many individuals may be tested to determine the probability of that specific genotype occurring. The likelihood ratio (LR) and the combined probability of exclusion for paternity (CPE_P) can be calculated from the allele frequencies (p and q) assigned to the experimentally determined STR profiles. The likelihood ratio is most often used in paternity cases, whereas the CPE is rarely used in cases of related individuals because of the small pool of included alleles. First, the child's genetic profile is compared to that of the

TABLE 18.1   Inheritance of Autosomal, Y Chromosome, X Chromosome, and Mitochondrial DNA Alleles

Inheritance	Autosomal Alleles	Y Alleles	mtDNA	X Alleles
Mother → Son	50%	N/A	100%	100%
Mother → Daughter	50%	N/A	100%	50%
Father → Son	50%	100%	0%	0%
Father → Daughter	50%	0%	0%	100%
Paternal Grandmother → Granddaughter	25%	N/A	0%	100%
Maternal Grandmother → Granddaughter	25%	N/A	100%	25%
Paternal Grandfather → Grandson	25%	100%	0%	0%

known parent. The alleles that can be assigned as to the known parent will be assigned using Punnett squares. The remaining allele, or both alleles, is referred to as the obligate allele(s). If the alleged parent and the child share an obligate allele at a locus, the individual cannot be excluded as a potential biological parent for that locus. In criminal paternity cases in which one or two alleles of the alleged parent do not match the profile of the child, the variation may be considered to arise from a mutational event. There must be an exclusion at three or more loci for an alleged parent to be excluded as a potential biological parent. An exclusion is recorded if the alleged parent does not share the obligate allele(s) with the child at more than two loci. No statistics will be recorded. If the alleged parent cannot be excluded (is included) from all interpretable loci, statistics will be utilized. However, the case study in this experiment is one of a recently reported three-loci exclusion (Sun et al., 2011) (Table 18.2).

The likelihood ratio is a comparison of the probabilities of two mutually exclusive hypotheses: the hypothesis of the prosecution, H_p, and the hypothesis of the defense, H_d. The likelihood ratio is most often used in paternity cases. The hypothesis of the prosecution concludes that the DNA matches because it is the suspect's DNA or DNA from the crime scene is the suspect's DNA, whereas the hypothesis of the defense concludes that the DNA just happens to match by coincidence and is instead a random match from some person in the population at large. The likelihood ratio is computed from the ratio of the H_p/H_d. The hypothesis of the prosecution is that the suspect committed the crime, so $H_p = 1$. The hypothesis of the defense considers the genotype rarity in the population or p^2 for a homozygote and $2pq$ for a heterozygote. A likelihood ratio greater than 1 supports the hypothesis of the prosecution, and a likelihood ratio of less than 1 supports the hypothesis of the defense. The likelihood ratio gets stronger as more loci are profiled (Table 18.3). The overall likelihood ratio is the product of the likelihood ratios computed for the individual loci:

$$LR = H_p/H_d = H_0/H_1 = 1/p^2 \text{ (homozygote) or } 1/2pq \text{ (heterozygote)}$$

(Alternatively, for related individuals in cases of paternity, the table at www.nfstc.org/pdi/Subject07/pdi_s07_m02_06_b.htm may be used.)

$$\text{Overall LR} = LR_1 * LR_2 * LR_3 \ldots LR_n$$

The combined probability of exclusion can be calculated from the following equations:

$$CPE_P = 1 - [1 - (PE_{Locus\ 1})] \times [1 - (PE_{Locus\ 2})] \ldots \times [1 - (PE_{Locus\ n})]$$

where, for homozygotes (frequencies of other alleles): $PE_{locus} = (1 - p)^2$.
And for heterozygotes (frequencies of other alleles):

$$PE_{locus} = [1 - (p + q)]^2$$

(For nonpaternity cases : $PE = 2pq + q^2$)

The CPE calculations are useful for determining the rarity of a profile observed in a mixture or other sample without assigning the alleles to a given individual or pairing them in cases of similar peak heights. The method is useful for estimating the portion of the population that has a genotype composed of at least one allele not observed in the profile.

TABLE 18.2   Data for 26 STR Loci for Sample Paternity Case

Locus	AF	C	M	GM
D3S1358	16, 17	17, 18	18, 18	17, 18
TH01	7, 9	9, 9	9, 9	6, 9
D21S11[b]	29, 32.2	31, 31.2	30, 31	26, 31
D18S51	14, 17	17, 19	13, 19	13, 13
D5S818	11, 13	11, 13	11, 11	11, 10
D13S317	8, 11	8, 8	8, 8	8, 8
D7S820	11, 12	10, 11	10, 10	10,13
D16S539	11, 12	10, 11	10, 12	12, 14
CSF1PO	11, 11	11, 12	12, 12	11, 12
vWA[b]	17, 18	16, 19	17, 19	17, 20
D8S1179[b]	15, 15	14, 15	15, 15	15, 15
TPOX	11, 11	11, 11	11, 11	9, 11
FGA	23, 24	23, 24	21, 23	18.2, 23
Penta D	9, 9	9, 12	9, 12	n/d
Penta E	16, 18	18, 20	12, 20	n/d
D2S1338	23, 24	19, 23	19, 23	n/d
D19S433	14, 15	14, 16.2	13, 16.2	n/d
D18S1364	13, 14	14, 15	14, 15	n/d
D12S391	17, 18	17, 19	19, 19	n/d
D13S325	18, 20	18, 21	20, 21	n/d
D6S1043	12, 14	14, 19	19, 20	n/d
D2S1772	21, 27	21, 22	21, 22	n/d
D11S2368	21, 22	22, 22	22, 22	n/d
D22-GATA 198B05	17, 21	17, 21	17, 22	n/d
D8S1132	20, 21	18, 20	18, 21	n/d
D7S3048	18, 25	22, 25	22, 25	n/d

AF = Alleged father
C = Child
M = Mother
n/d = no data
GM = maternal grandmother
26 STR loci for sample paternity case (Adapted from *Sun et. al., 2011*) for AF, C, and M trio. The first 13 loci (italics) are the CODIS loci. *STR typing kits used included PowerPlex 16 (Promega), Identifiler Plus (Applied Biosystems) and STRTyper 10 (Codon).*

TABLE 18.3 Scoring Likelihood Ratios

Likelihood Ratio	Degree of Support
1 to 10	Limited support
10 to 100	Moderate support
100 to 1000	Strong support
>1000	Very strong support

Parentage trios of a mother, child, and alleged father have far fewer alleles among them at given loci than the entire population. To demonstrate, the number of genotypes possible for a population with a given number of alleles can be calculated. For example, for TH01, there are eight alleles (5, 6, 7, 8, 8.3, 9, 9.3, 10), so

$$\text{Total number of genotypes} = (n(n + 1))/2$$

At TH01, there are $(8(8 + 1))/2$ or 36 genotypes (8 homozygous and 28 heterozygous). However, in a parentage trio, if there are three alleles (e.g., 5, 7, 8), there are $(3(3+1))/2$ or 6 genotypes.

A random match probability is not the estimated frequency at which an STR profile would be expected to occur in a population, the theoretical chance that a person at random in a given population will have DNA profile in question, the frequency that another individual is guilty or left the biological material at the crime scene, the chance that the defendant is not guilty, or the chance someone else in reality would have same genotype.

This lab is based on understanding the principles of Mendelian genetics as applied to STR loci including X and Y markers. In this lab you are presented data from a sample case (Sun et al., 2011) that includes the DNA profiles for the alleged father (AF), mother (M), and child (C). In the first part, you will use Punnett squares to evaluate whether the parent genotypes could produce the child genotype shown for each locus. You will calculate the number of genotypes possible for a given parentage trio as compared to the population using Tables 18.2 and 18.4. You will compute the LR and CPE_P for the child's data for the 13 CODIS loci and interpret the results of this and other scenarios. Finally, you will assign a partial profile that can be deduced from the profiles of the mother (M) and grandmother (GM) in Table 18.2 for the missing grandfather (GF).

PROCEDURE

Part A: Paternity and Genotype Calculations

1. Draw Punnett squares for each of the given STR loci in Table 18.2 using the parental alleles given.
2. Compute the number of genotypes possible for the 13 CODIS loci using Table 18.4. Compare this to how many genotypes are possible using the profiles of the two parents. Comment on the numbers.

TABLE 18.4   Allele Frequencies of 13 CODIS Loci (Budowle et al., 1999)

D3S1358	African American (N= 210)	Caucasian (N= 203)	Hispanic (N= 209)
<12	0.00476	0.00000	0.00000
12	0.00238	0.00000	0.00000
13	0.01190	0.00246	0.00239
14	0.12143	0.14039	0.07895
15	0.29048	0.24631	0.42584
15.2	0.00000	0.00000	0.00000
16	0.30714	0.23153	0.26555
17	0.20000	0.21182	0.12679
18	0.05476	0.16256	0.08373
19	0.00476	0.00493	0.01435
>19	0.00238	0.00000	0.00239
VWA	**African American (N=180)**	**Caucasian (N=196)**	**Hispanic (N=203)**
11	0.00253	0.00000	0.00246
13	0.00556	0.00510	0.00000
14	0.06667	0.10204	0.06158
15	0.23611	0.11224	0.07635
16	0.26944	0.20153	0.35961
17	0.18333	0.26276	0.22167
18	0.13611	0.22194	0.19458
19	0.07222	0.08418	0.07143
20	0.02778	0.01020	0.01232
21	0.00000	0.00000	0.00000
FGA	**African American (N=180)**	**Caucasian (N=196)**	**Hispanic (N=203)**
<18	0.0028	0.0000	0.0000
18	0.0083	0.0306	0.0025
18.2	0.0083	0.0000	0.0000
19	0.0528	0.0561	0.0788
19.2	0.0003	0.0000	0.0000
20	0.0722	0.1454	0.0714
20.2	0.0000	0.0026	0.0025
21	0.1250	0.1735	0.1305

TABLE 18.4 Allele Frequencies of 13 CODIS Loci (Budowle et al., 1999)—cont'd

D3S1358	African American (N = 210)	Caucasian (N = 203)	Hispanic (N = 209)
21.2	0.0000	0.0000	0.0025
22	0.2250	0.1888	0.1773
22.2	0.0056	0.0102	0.0049
22.3	0.0000	0.0000	0.0000
23	0.1250	0.1582	0.1404
23.2	0.0000	0.0000	0.0074
24	0.1861	0.1378	0.1256
24.2	0.0000	0.0000	0.0000
24.3	0.0000	0.0000	0.0000
25	0.1000	0.0689	0.1379
26	0.0361	0.0177	0.0837
27	0.0222	0.0102	0.0320
28	0.0167	0.0000	0.0025
29	0.0056	0.0000	0.0000
30	0.0028	0.0000	0.0000
>30	0.0028		0.0000
D8S1179	**African American (N=180)**	**Caucasian (N=196)**	**Hispanic (N=203)**
<9	0.00278	0.01786	0.00246
9	0.00556	0.01020	0.00246
10	0.02500	0.10204	0.09360
11	0.03611	0.05867	0.06158
12	0.10833	0.14541	0.12069
13	0.22222	0.33929	0.35512
14	0.33333	0.20153	0.24631
15	0.21389	0.10969	0.11576
16	0.04444	0.01276	0.02463
17	0.00833	0.00255	0.00739
18	0.00000	0.00000	0.00000
D21S11	**African American (N=179)**	**Caucasian (N=196)**	**Hispanic (N=203)**
24.2	0.00279	0.00510	0.00246
24.3	0.00000	0.00000	0.00000

(Continued)

TABLE 18.4   Allele Frequencies of 13 CODIS Loci (Budowle et al., 1999)—cont'd

D3S1358	African American (N= 210)	Caucasian (N= 203)	Hispanic (N= 209)
26	0.00279	0.00000	0.00000
27	0.06145	0.04592	0.00985
28	0.21508	0.16582	0.06897
29	0.18994	0.18112	0.20443
29.2	0.00279	0.00000	0.00246
30	0.17877	0.23214	0.33005
30.2	0.00838	0.03827	0.03202
30.3	0.00000	0.00000	0.00000
31	0.09218	0.07143	0.06897
31.2	0.07542	0.09949	0.08621
32	0.00838	0.01531	0.01232
32.1	0.00000	0.00000	0.00000
32.2	0.06983	0.11224	0.13547
33	0.00838	0.00000	0.00000
33.2	0.03352	0.03061	0.04187
34	0.00838	0.00000	0.00000
34.2	0.00279	0.00000	0.00493
35	0.02793	0.00000	0.00000
35.2	0.00000	0.00255	0.00000
36	0.00559	0.00000	0.00000
>36	0.00559	0.00000	0.00000
D18S51	**African American (N=180)**	**Caucasian (N=196)**	**Hispanic (N=203)**
<11	0.00556	0.01276	0.00493
11	0.00556	0.01276	0.01232
12	0.05833	0.12755	0.10591
13	0.05556	0.12245	0.16995
13.2	0.00556	0.00000	0.00000
14	0.06389	0.17347	0.16995
14.2	0.00000	0.00000	0.00000
15	0.16667	0.12755	0.13793
15.2	0.00000	0.00000	0.00000
16	0.18889	0.10714	0.11576

TABLE 18.4 Allele Frequencies of 13 CODIS Loci (Budowle et al., 1999)—cont'd

D3S1358	African American (N= 210)	Caucasian (N= 203)	Hispanic (N= 209)
17	0.16389	0.15561	0.13793
18	0.13056	0.09184	0.05172
19	0.07778	0.03571	0.03695
20	0.05556	0.02551	0.01724
21	0.01111	0.00510	0.01970
21.2	0.00000	0.00000	0.00000
22	0.00556	0.00255	0.00739
>22	0.00556	0.00000	0.01232
D5S818	**African American (N=180)**	**Caucasian (N=195)**	**Hispanic (N=203)**
7	0.00278	0.00000	0.06158
8	0.05000	0.00000	0.00246
9	0.01389	0.03077	0.05419
10	0.06389	0.04872	0.06650
11	0.26111	0.41026	0.42118
12	0.35556	0.35385	0.29064
13	0.24444	0.14615	0.09606
14	0.00556	0.00769	0.00493
15	0.00000	0.00256	0.00246
>15	0.00278	0.00000	0.00000
D13S317	**African American (N=179)**	**Caucasian (N=196)**	**Hispanic (N=203)**
7	0.00000	0.00000	0.00000
8	0.03631	0.09949	0.06650
9	0.02793	0.07653	0.21921
10	0.05028	0.05102	0.10099
11	0.23743	0.31888	0.20197
12	0.48324	0.30867	0.21675
13	0.12570	0.10969	0.13793
14	0.03631	0.03571	0.05665
15	0.00279	0.00000	0.00000
D7S820	**African American (N=210)**	**Caucasian (N=203)**	**Hispanic (N=209)**
6	0.00000	0.00246	0.00239
7	0.00714	0.01742	0.02153

(Continued)

TABLE 18.4   Allele Frequencies of 13 CODIS Loci (Budowle et al., 1999)—cont'd

D3S1358	African American (N= 210)	Caucasian (N= 203)	Hispanic (N= 209)
8	0.17381	0.16256	0.09809
9	0.15714	0.14778	0.04785
10	0.32381	0.29064	0.30622
10.1	0.00000	0.00000	0.00000
11	0.22381	0.20197	0.28947
11.3	0.00000	0.00000	0.00000
12	0.09048	0.14039	0.19139
13	0.01905	0.02956	0.03828
14	0.00476	0.00739	0.00478
CSF1PO	**African American (N=210)**	**Caucasian (N=203)**	**Hispanic (N=209)**
6	0.00000	0.00000	0.00000
7	0.04286	0.00246	0.00239
8	0.08571	0.00493	0.00000
9	0.03333	0.01970	0.00718
10	0.27143	0.25369	0.25359
10.1	0.00000	0.00246	0.00000
11	0.20476	0.30049	0.26666
11.3	0.30000	0.32512	0.39234
12	0.05476	0.07143	0.06459
13	0.00714	0.01478	0.00957
14	0.00000	0.00493	0.00478
TPOX	**African American (N=209)**	**Caucasian (N=203)**	**Hispanic (N=209)**
6	0.08612	0.00000	0.00478
7	0.02153	0.00246	0.00478
8	0.36842	0.54433	0.55502
9	0.18182	0.12315	0.03349
20	0.09330	0.03695	0.03349
11	0.22488	0.25369	0.27273
12	0.02392	0.03941	0.09330
13	0.00000	0.00000	0.00239

TABLE 18.4   Allele Frequencies of 13 CODIS Loci (Budowle et al., 1999)—cont'd

D3S1358	African American (N= 210)	Caucasian (N= 203)	Hispanic (N= 209)
TH01	**African American (N=210)**	**Caucasian (N=203)**	**Hispanic (N=209)**
5	0.00000	0.00000	0.00239
6	0.10952	0.22660	0.23206
7	0.44048	0.17241	0.33732
8	0.18571	0.12562	0.08134
8.3	0.00000	0.00246	0.00000
9	0.14524	0.16502	0.10287
9.3	0.10476	0.30542	0.24163
10	0.01429	0.00246	0.00239
D16S539	**African American (N=209)**	**Caucasian (N=202)**	**Hispanic (N=208)**
8	0.03589	0.01980	0.01683
9	0.19856	0.10396	0.07933
10	0.11005	0.06683	0.17308
11	0.29426	0.27228	0.31490
12	0.18660	0.33911	0.28606
13	0.16507	0.16337	0.10337
14	0.00957	0.03218	0.02404
15	0.00000	0.00248	0.00240

Data taken from Budowle et al., 1999.

3. Using the population database in Table 18.4, assign the allele frequencies for each of the child's alleles for the 13 CODIS loci in Table 18.2.
4. Compute the CPE_P and LR for each of the child's 13 CODIS loci and the overall CPE_P and LR.
5. Interpret the result of your analysis? Do the data support the conclusion that the child is the biological offspring of these parents based on Table 18.3? Report the result as included, inconclusive, or excluded. What do you think is the meaning of the "b" superscript notation in the table?

Part B: Reverse Parentage Questions (Show Your Work Using Punnett Squares)

1. If the genotypes of a mother and several children are known, it is often possible to unambiguously predict the genotype of the father. If the mother has a genotype of 7, 8 and her children have genotypes of 5, 8; 7, 8; and 8, 8 at the TH01 locus, determine the genotype of the father.

2. Sometimes only one allele of the father can be predicted when the genotypes of a mother and several children are known. The genotype of the mother is 16, 17 at the D3S1358 locus. The genotypes of her daughters are 16, 18 and 17, 18. Which allele was inherited from the father? Is the father homozygous or heterozygous?

3. If four children share the same biological parents and the genotypes of the children at D3S1358 are 12, 13; 12, 18; and 17, 18, what are the genotypes of the two parents? Can you assign the genotypes to the parents?

4. Sometimes the parental genotypes cannot be predicted unambiguously from the genotypes of their children. If the genotypes of the three offspring of two parents at CSF1PO are 8, 9; 8, 10; and 8, 10, what are the possible parental genotypes?

Part C: Missing Persons Questions (Show Your Work Using Punnett Squares.)

Consider the mother's and grandmother's (GM) profile in Table 18.2. If her father (the grandfather, GF) is missing, predict the partial profile of the grandfather (GF).

References

Budowle, B., Moretti, T.R., Baumstark, A.L., Defenbaugh, D.A., Keys, K.M., 1999. Population data on the thirteen CODIS core short tandem repeat loci in African Americans, U.S. Caucasians, Hispanics, Bahamians, Jamaicans, and Trinidadians. J. Forensic Sci. 44, 1277–1286.

Sun, H.-y., Li, H.-x., Zeng, X.-p., Ren, Z., Chen, W.-j., 2011. A paternity case with mutations at three CODIS core STR loci. Forensic Sci. Int. Genet. http://dx.doi.org/10.1016/j.fsigen.2011.05.006.

Low Copy Number Stochastic Results

OBJECTIVE

To examine the short tandem repeat (STR) profile for a low copy number deoxyribonucleic acid (DNA) sample derived by from serial dilution of a standard DNA template to exemplify of situations where peak imbalance results in only one detectable allele from a heterozygous pair.

SAFETY

Handle the SYBR Green I reaction mix carefully; SYBR Green I is an intercalating agent but is not noted to be mutagenic. If you spill any chemicals on your person, wash them off immediately with soap and water.

The real-time PCR instrument is very sensitive to background noise. Do not touch the plate without gloved hands. To avoid contamination, wear gloves when handling the samples.

MATERIALS

1. BioRad iQ 96-Well PCR Plates
2. Microseal "B" film
3. 2x iQ SYBR Green SuperMix
4. Nuclease-free water
5. Extracted DNA template (known 1- to 10-ng range) or K562 (or other desired template) DNA
6. Selected PowerPlex 16 (Promega) primers (diluted to 5 μM) (IDT, custom, 25-nmole, standard desalting) for D7S820, D5S818 and Penta D (Table 19.1)
7. BioRad iQ5 RT-PCR instrument with BioRad iQ5 software v. 2.0 or traditional PCR thermocycler
8. 1.5-mL microcentrifuge tubes

TABLE 19.1 Expected Allele Determinations for Standard DNA Templates for PowerPlex 16 (Promega) Primers

Locus	Standard DNA Template			Size Range	Forward Primer	T_m (°C)	Reverse Primer	T_m (°C)	Dye Label in Kit
	K562	9947A	9948						
Amelogenin	X, X	X, X	X, Y	106, 112	CCCTGGGC TCTGTAAAGAA	54.2	ATCAGAGCT TAAACTG GGAAGCTG	57.0	TMR
D3S1358	16, 16	15, 14	17, 15	115-147	ACTGCAGTC CAATCTGGGT	56.4	ATGAAATC AACAGAGGC TTGC	53.7	Fluorescein
D5S818	12, 11	11, 11	13, 11	119-155	GGTGATTTTCC TCTTTGGTATCC	53.5	AGCCACAGTTT ACAACATTTG TATCT	55.1	JOE
vWA	16, 16	18, 17	17, 17	123-171	GCCCTAGTGGA TGATAAGAA TAATCAGTATGTG	58.1	GGACAGATG ATAAATACATA GGATGGATGG	56.5	TMR
TH01	9.3, 9.3	9.3, 8	9.3, 6	156-195	GTGATTCCCA TTGGCCTGTTC	56.5	ATTCCTGTGG GCTGAAAA GCTC	57.8	Fluorescein
D13S317	8, 8	11, 11	11, 11	169-201	ATTACAGAAG TCTGGGATG TGGAGA	59.0	GGCAGCCCA AAAAGACAGA	55.8	JOE
D8S1179	12, 12	13, 13	13, 12	203-247	ATTGCAACTTA TATGTATTTT TGTATTTCATG	52.5	ACCAAATTG TGTTCATGAG TATAGTTTC	54.0	TMR
D21S11	31, 30, 29	30, 30	30, 29	203-259	ATATGTGAGTC AATTCCCCAAG	52.5	TGTATTAGT CAATGTTCT CCAGAGAC	54.2	Fluorescein
D7S820	11, 9	11, 10	11, 11	215-247	ATGTTGGTCA GGCTGACTATG	54.6	GATTCCACA TTTATCCT CATTGAC	51.9	JOE
TPOX	9, 8	8, 8	9, 8	262-290	GCACAGAA CAGGCACTTAGG	56.0	CGCTCAAACG TGAGGTTG	54.2	TMR

Locus				Range	Sequence	Tm	Sequence	Tm	Dye
D16S539	12, 11	12, 11	11, 11	264-304	GGGGGTCTAA GAGCTTGTAA AAAG	55.5	GTTTGTGT GTGCATCTGT AAGCATGTATC	58.2	JOE
D18S51	16, 15	19, 15	18, 15	290-366	TTCTTGAGC CCAGAAGGTTA	53.4	ATTCTACCAG CAACAACAC AAATAAAC	54.6	Fluorescein
CSF1PO	10, 9	12, 10	12, 11, 10	321-357	CCGGAGGTAAA GGTGTCTTA AAGT	56.4	ATTTCCTG TGTCAG ACCCTGTT	56.3	JOE
FGA	24, 21	24, 23	26, 24	322-444	GGCTGCAGGG CATAACATTA	55.5	ATTCTATGAC TTTGCGCTT CAGGA	56.5	TMR
Penta D	13, 9	12, 12	12, 8	376-441	GAAGGTCGAA GCTGAAGTG	53.3	ATTAGAA TTCTTTAATCT GGACACAAG	51.7	JOE
Penta E	14, 5	13, 12	11, 11	379-474	ATTACCAA CATGAAAGGG TACCAATA	54.4	TGGGTTATTAA TTGAGAAAA CTCCTTACAATTT	55.8	Fluorescein

9. Freezer
10. Various variable volume micropipettes (e.g., 0.5 to 10 μL, 10 to 100 μL, 100 to 1000 μL)
11. Autoclaved pipette tips for pipettes
12. Sterile microcentrifuge tubes or PCR tubes for preparing samples
13. Gloves, lab coats, and goggles
14. 6X loading dye for gel
15. Heating block or water bath
16. Agarose
17. 1X TAE buffer
18. Gel boxes and power supply
19. 10,000 to 100 bp DNA ladder or other ladder
20. Microwave or hotplate
21. 100X SYBR Green I

BACKGROUND

Commercial multiplex STR DNA typing kits used in crime laboratories work optimally under a narrow range of conditions. Adding the standard quantity of DNA, 1 to 2.5 ng, is time and cost effective, as the kits have been optimized for this condition. The DNA Advisory Board Standard 9.3 requires human-specific quantitation so that the user can supply the appropriate level of DNA in the PCR amplification. So what happens if no DNA is detected? For example, a DNA quantification test is run, and fluorescence spectroscopy, quantitative PCR (qPCR), QuantiBlot, or Quantifiler is negative for detectable levels of DNA in the samples. DNA fingerprinting has been performed from fingerprints (van Oorschot and Jones, 1997) and with as little DNA as that extracted from a single cell (Findlay et al., 1997).

Low copy number (LCN) is defined as less than 100 pg of DNA or the equivalent of the DNA contained in 15 to 17 diploid cells (Gill, 2001). This is the stochastic (random selection or statistical sampling of the chromosomes) threshold beyond which PCR amplification is not reliable. LCN DNA may also be referred to as touch DNA, trace DNA, and low-level DNA. However, LCN samples may be present in low-level contributors in mixtures even with a 1 ng quantity of human DNA. The results of LCN samples using multiplex STR DNA typing kits are heterozygous peak imbalance (difference in relative fluorescence units of peak heights for a heterozygote at a given locus), allele dropout (one or more alleles does not amplify), locus-to-locus imbalance (difference in relative fluorescence units of peak heights for alleles at different loci), and even allele drop-in. As the results are random, they are referred to as stochastic effects.

The stochastic effect is a random selection or statistical sampling of the chromosomes, a threshold beyond which PCR amplification is not reliable. Some stochastic effects include heterozygous peak imbalance, allele drop-out, locus-to-locus imbalance, and allele drop-in. Heterozygous peak imbalance results in a difference in relative fluorescence units of peak heights for a heterozygote at a given locus. Allele drop-out is characterized by one or more alleles which fails to amplify. Locus-to-locus imbalance is a term to describe a difference in relative fluorescence units of peak heights for alleles at different loci. Allele drop-in refers to extra alleles observed in a profile.

The commercial multiplex STR DNA typing kits typically have been optimized to give a full profile after 28 PCR cycles with at least 125 pg input DNA. The sensitivity is due to a combination of the optimized fluorescent dye characteristics, laser excitation source, detection system, and optimized quantity of PCR primers, dNTPs, magnesium chloride, potassium chloride, bovine serum albumin, and DNA polymerase. By increasing the number of PCR cycles from 28 to 34 or even 42 cycles, the DNA typing kits are often more sensitive than the quantification detection methods (Krenke et al., 2002, Whitaker et al., 2001, Wallin et al., 1998). Likewise, a simple, one-target PCR reaction can detect lower quantities of DNA, as the number of cycles is increased. The important issue is reliability. A DNA sample for which a given locus is expected to be a heterozygote (or homozygote) must be typed and be able to be called as the heterozygote (or homozygote) using a DNA typing method. The detection limit is defined as the minimum quantity of sample that can be measured with reasonable scientific certainty. With LCN samples, the DNA will be typed using three independent PCR reactions (replicates), the present alleles will be assigned and a consensus profile will be developed based on the replicate consistent results (Gill, 2000, Gill et al., 2000). An allele cannot be considered to be real unless it is present in at least two amplifications. An extremely sterile environment is important to avoid contamination from laboratory personnel or individuals with access to the lab. Other methods to increase sensitivity in methods for evaluating LCN profiles are increasing the injection time on the capillary electrophoresis, removing salt from samples, concentrating PCR product prior to loading on capillary electrophoresis, using highest yield DNA extraction methods, increasing the quantity of DNA polymerase, or using mini-STRs or mtDNA sequencing as an alternative to standard multiplex conditions.

In this lab, you will make serial dilutions of a DNA standard sample, K562, and evaluate which loci are heterozygotes at all loci by evaluating the quantity of DNA produced and the ratio of DNA amplified for each allele. The outcomes will be evaluated either using PCR and agarose gel electrophoresis or using a commercial DNA typing kit and capillary electrophoresis (the peak height ratio can be evaluated for the concentration in which a set of peak height ratio values fall below 60%). The Promega PowerPlex16 loci that are heterozygotes for K562 are D5S818, D7S820, TPOX, D16S539, D18S51, CSF1PO, FGA, Penta D, and Penta E (see Table 19.1). D7S820, D5S818, and Penta D will be amplified in this lab using the primers shown in Table 19.1.

Sample data for real-time PCR amplification, melt curve, and agarose gels are shown in Figures 19.1, 19.2, and 19.3, respectively. In Figure 19.2, the melt curve clearly shows the presence of three size range amplicons, as there are three resolvable peaks. In a 2% agarose gel (Figure 19.3), the amplicons were separated so that all heterozygotes are observed down to at least 100 pg. Below this concentration, stochastic effects are observed. In these sample data, no amplification was observed with the zero and one picogram concentrations. The expected amplicon sizes for K562 are 135 and 139 bp at D5S818, 227 and 235 bp at D7S820, and 417 and 425 bp at Penta D.

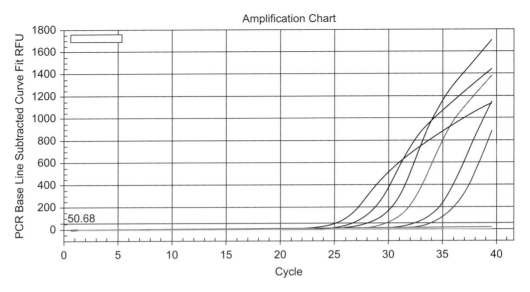

FIGURE 19.1 Exponential amplification curves that show the results of 40 cycles of RT-PCR using PowerPlex 16 (Promega) D7S820, D5S818, and Penta D primers depicted in Table 19.1. Curves of K562 DNA as detected by SYBR green I and a flat line for the negative control.

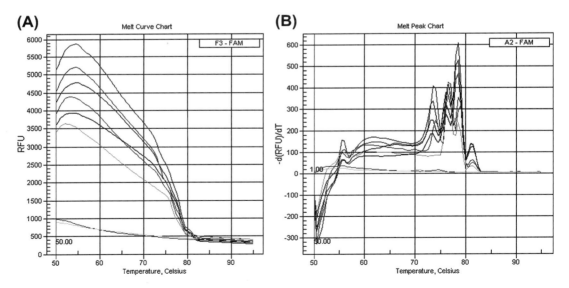

FIGURE 19.2 A. Melting curve plot from 50° to 95° C for the three amplicons produced from K562 DNA (A). B. First derivative plot of the melting curve from 50° to 95° C showing the predominant peaks at 73.5° C (D5S818), 76° C (D7S820), and 78.5° C (Penta D) for the amplicons produced from the K562 DNA (B).

FIGURE 19.3 Sample 2% agarose gel of PCR of K562 DNA using PowerPlex16 primers D7S820, D5S818, and Penta D. Lanes 1 and 14 contain the Fisher BioReagents* exACTGene* DNA Ladders 2kb DNA Ladder (sizes indicated). Lanes 2 through 9 contain amplicons produced using 1000 pg, 500 pg, 100 pg, 50 pg, 10 pg, 5 pg, 1 pg, and 0 pg, respectively. Lanes 10 through 13 are blank.

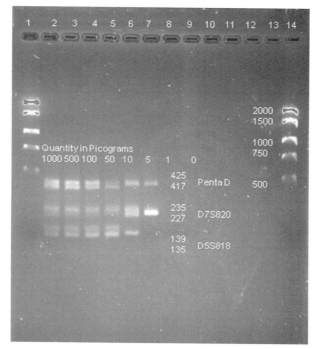

PROCEDURE

Week 1 (PCR)

1. Perform a serial dilution (1 ng to 1 pg) of a control sample of K562 (or other provided or previously quantified) DNA.
2. Pipette the following into two separate 1.5-mL microcentrifuge tubes for the extracted DNA samples:
   12.5 µL of 2x iQ SYBR Green SuperMix
   1 µL of 5 µM of each of three forward primers
   1 µL of 5 µM of each of three reverse primers
   5.5 µL of nuclease-free water
   <u>1 µL of extracted template DNA (0 pg to 1 ng)</u>
   25 µL Total Reaction Volume
3. Prepare one negative control sample (substitute nuclease-free water for template DNA, so add 10.5 µL nuclease-free water) (individually or as a class).
4. Mix samples by pipetting up and down gently so as not to introduce bubbles. Centrifuge to remove bubbles.
5. Pipet your samples into the 96-well plate; carefully record wells used. Pipet the sample into the bottom of the well, and do not release the pipette until withdrawn from the sample so as to not produce bubbles.

6. When all samples have been loaded to the 96-well plate, use the plastic sealing device to seal Microseal plastic covers over all samples. Alternatively, use thin-walled PCR tubes.
7. Place all the tubes in the PCR Thermal Cycler, and begin the preprogrammed cycles listed below:
    95° C for 3 minutes
       95° C for 30 seconds
       60° C for 30 seconds
       72° C for 45 seconds (60 seconds per kb)
    Repeat the preceding three-step cycle 40 times.

Week 2 (Agarose Gel Electrophoresis)

(Alternative: Capillary electrophoresis may be used if dye-labeled primers are purchased. Polyacrylamide gel electrophoresis may also be used to separate the PCR products.)

1. Prepare a 2% agarose gel. Combine 1 g of agarose and 50 mL of 1x TAE buffer (which offers better resolution of DNA fragments than TBE buffer) in an Erlenmeyer flask covered with plastic wrap that has been punctured with a small hole.
2. Heat in the microwave for 40 seconds or until it boils and agarose is fully dissolved.
3. Remove from the microwave.
4. Wait for the agarose-TAE solution to cool to the touch, then pour it into the gel box positioned so that a rectangular box with four sides is created. Add a 14-well comb.
5. Once the gel is set (translucent blue color), rotate the gel so that the wells are positioned nearest the negative pole (black).
6. Fill gel box with 1x TAE buffer to cover the gel.
7. Remove the comb from the gel.
8. To prepare samples for the gel, add 3 μL of the 6x loading dye in the PCR tubes or 96-well plate.
9. Load 12 μL of the amplified samples-loading buffer to separate wells carefully by pipetting. Do not release the plunger until the pipette is removed from the gel so as to not draw up the samples.
10. To prepare a ladder for the gel, combine 8 μL of the DNA Logic ladder (or other ladder) with 1 μL of 100x SYBR Green I dye and 1 μL of 6X loading dye, and combine 4 μL of the Lonza 20 bp ladder with 1 μL of 100x SYBR Green I dye and 1 μL of the 6X gel loading buffer. (Refer to the manufacturer's instructions for other ladders.)
11. Load the entire portion of the two ladders to two different lanes on opposite sides of the gel.
12. Place the safety cover on the gel box to run the sample toward the positive (red) electrode. Connect and turn on the power supply.
13. Run the gel for 60 minutes at no more than 15 V/cm (or 150 V for 10 cm gel).
14. Turn off the power supply and unplug it. Remove the gel safety cover.
15. While wearing gloves, remove the gel from the casting plate, place it on a UV light box, and take a picture using a digital photodocumentation system with a SYBR filter.

QUESTIONS

1. Which loci had relative peak heights above 150 RFU or the detection limits? Which were not detectable on the gel or capillary electrophoresis electropherogram?
2. Which alleles were not detectable? At what concentration were they not detectable?
3. What were the expected and observed T_m values for the amplicons (if the melting curve was performed)? Give reasons for any deviations.
4. From the gel, estimate the size of the amplicon produced by measuring the distances of the ladder fragments and the amplicon fragment(s) and plotting log base pairs versus the distance traveled for the ladder and using the line equation and distances migrated by the unknowns to compute the size(s) of the amplicon(s).
5. Comment on the production of the expected amplicons in the LCN triplex reaction.

References

Findlay, I., Taylor, A., Quirke, P., Frazier, R., Urquhart, A., 1997. DNA fingerprinting from single cells. Nature 389, 555–556.

Gill, P., 2000. An Investigation to the rigor of interpretation rules for STRs derived from less than 100 pg of DNA. Forensic Science International 112, 37–40.

Gill, P., 2001. Application of low copy number DNA profiling. Croatian Medical Journal 42, 229–232.

Gill, P., Whitaker, J., Flaxman, C., Brown, N., Buckleton, J., 2000. An investigation of the rigor of interpretation rules for STRs derived from less than 100 pg of DNA. Forensic Sci. Int. 112 (2000), 17–40.

Krenke, B.E., Tereba, A., Anderson, S.J., Buel, E., Culhane, S., Finis, C.J., Tomsey, C.S., Zachetti, J.M., Masibay, A., Rabbach, D.R., Amiott, E.A., Sprecher, C.J., 2002. Validation of a 16-locus fluorescent multiplex system. J. Forensic Sci. 47, 773–785.

van Oorschot, R.A., Jones, M.K., 1997. DNA fingerprints from fingerprints. Nature 387 (1997), 767.

Wallin, J.M., Buoncristiani, M.R., Lazaruk, K.D., Fildes, N., Holt, C.L., Walsh, P.S., 1998. TWGDAM validation of the AmpFlSTR blue PCR amplification kit for forensic casework analysis. J. Forensic Sci. 43, 854–870.

Whitaker, J.P., Cotton, E.A., Gill, P., 2001. A comparison of the characteristics of profiles produced with the AMPFlSTR SGM Plus multiplex system for both standard and low copy number (LCN) STR DNA analysis. Forensic Sci. Int. 123, 215–223.

Using *in Silico* Methods to Construct a Short-Tandem Repeat (STR) Deoxyribonucleic Acid (DNA) Sequence for Cloning

OBJECTIVE

To learn how to use the web servers National Center for Biotechnology Information (NCBI) to retrieve a STR sequence, Integrated DNA Technologies' (IDT) OligoAnalyzer tools to predict the complementary DNA strand, and New England (NE) Biolabs' website to look up the sequences cleaved by restriction enzymes and add those sequences to an upstream and downstream region of the DNA in order to insert the synthetic DNA into the pET-15b plasmid used to clone the DNA in *Escherichia coli* bacteria.

SAFETY

No special safety precautions. This is a computer-based *in silico* lab.

MATERIALS

Computer with an Internet connection

BACKGROUND

In a recent paper, forgery of STR DNA through cloning the sequence in a plasmid and inserting it into a bacterium was highlighted (Frumpkin et al. 2010). In this experiment, modern computational tools will be used to obtain a nucleotide sequence containing the STR of interest from the NCBI database. The complementary strand of the DNA sequence of interest will be computed, and the plasmid map for the pET-15b cloning/expression plasmid will be obtained from the manufacturer's website. Using the website for New England Biolabs, the DNA sequences cleaved by the restriction enzymes found in the multiple cloning site region of the plasmid will be identified. The double-stranded DNA sequence will be prepared for cloning by adding the restriction enzyme sequences to the regions upstream and downstream of the PowerPlex 16 primer region (Table 20.1). The prepared DNA sequence could be purchased or produced by PCR amplification from a template. By using the restriction enzymes to cut the DNA sequence with the primer region intact and the restriction sites added and the plasmid, the DNA can be inserted into the plasmid and ligated using DNA ligase in the lab.

Specifically, the nucleotide sequence containing the STR and primer binding region will be obtained from NCBI using the accession number (Table 20.1) to obtain the nucleotide sequence, Integrated DNA Technologies' OligoAnalyzer DNA tools will be used to predict the complementary DNA strand, the plasmid map will be obtained from EMD Biosciences' website, and the cleavage sequences of the restriction enzymes will be obtained from New England Biolabs' website for the restriction enzymes contained in the cloning/expression region of the plasmid. The restriction enzyme sequences will be selectively added to the foreign piece of DNA upstream and downstream of the 5' and 3' primer sequences, respectively, in order to insert the DNA into the plasmid and code for the STR of interest.

Plasmids can be used to insert DNA into bacterial cells using a method called transformation. Restriction enzymes can be used to cut DNA in specific sequence locations leaving either a blunt end with (clean cut) or "sticky" end with a few base overhangs on one strand in a reaction called a digest. DNA ligase can be used to catalyze covalent bond formation between neighboring DNA bases previously cut in the plasmid or of the foreign DNA with the plasmid. Once inside the cell, the plasmid will be replicated, or amplified, by the bacteria using DNA polymerase. The plasmids containing the DNA sequence of interest could be extracted from the bacteria and planted at a crime scene, implicating a person with the STR allele(s) with the crime.

PROCEDURE

Part A: Obtaining a DNA Sequence from NCBI

1. Access www.ncbi.nlm.nih.gov.
2. Select Search: Nucleotide for one of the accession numbers listed in Table 20.1 noted to contain an STR sequence. Record the accession number STR you select.
3. Use the edit:find feature of the browser and enter part of the primer sequence for a locus of interest. Continue to scroll through the STR repeats (by clicking "Next") to the STR allele of interest. Copy and paste the STR and flanking regions (including at least PowerPlex16 primer binding regions upstream and downstream) into the Notepad accessory or Microsoft Word. If a different STR allele is desired, the number of STR repeats may be added or deleted.

Part B: Using IDT OligoAnalyzer Tools to Create the DNA Complement

1. Access www.idtdna.com/Home/Home.aspx.

TABLE 20.1 NCBI Accession Numbers for Loci Probed by PowerPlex16 Kits with Primer Sequences

Locus	NCBI Accession Number	Forward Primer	Reverse Primer
Amelogenin		CCCTGGGCTCTCTGTAAAGAA	ATCAGAGCTTAAACTGGGAAGCTG
D3S1358	NT_005997	ACTGCAGTCCAATCTGGGT	ATGAAATCAACAGAGGCTTGC
D5S818	AC008512 or G08446	GGTGATTTTCCTCTTTGGTATCC	AGCCACAGTTTACAACATTTGTATCT
vWA	M25858	GCCCTAGTGGATGATAAGAATAATCAGTATGTG	GGACAGATGATAAATACATAGGATGGATGG
TH01	D00269	GTGATTCCCATTGGCCTGTTC	ATTCCTGTGGGCTGAAAAGCTC
D13S317	AL353628 or G09017	ATTACAGAAGTCTGGGATGTGGAGGA	GGCAGCCCAAAAAGACAGA
D8S1179	AF216671 or G08710	**ATT**GCAACTTATATGTATTTTGTATTTCATG	ACCAAATTGTGTTCATGAGTATAGTTTC
D21S11	AP000433	ATATGTGAGTCAATTCCCCAAG	TGTATTAGTCAATGTTCTCCAGAGAC
D7S820	AC004848 or G08616	ATGTTGGTCAGGCTGACTATG	GATTCCACATTTATCCTCATTGAC
TPOX	M68651	GCACAGAACAGGCACTTAGG	CGCTCAAACGTGAGGTTG
D16S539	AC024591 or G07925	GGGGGTCTAAGAGCTTGTAAAAAG	GTTTGTGTGTGCATCTGTAAGCATGTATC
D18S51	AP001534 or L18333	TTCTTGAGCCCAGAAGGTTA	ATTCTACCAGCAACAACACAAATAAAC
CSF1PO	X14720 or U63963	CCGGAGGTAAAGGTGTCTTAAAGT	ATTTCCTGTGTCAGACCCTGTT
FGA	M64982	GGCTGCAGGGCATAACATTA	**A**TTCTATGACTTTGCGCTTCAGGA
Penta D	AP001752	GAAGGTCGAAGCTGAAGTG	ATTAGAATTCTTTAATCTGGACACAAG
Penta E	AC0270024	ATTACCAACATGAAAGGGTACCAATA	TGGGTTATTAATTGAGAAAACTCCTTACAATTT

2. Select OligoAnalyzer 3.0.
3. Paste PowerPlex 16 primer region with STR and flanking regions of DNA sequence into the box.
4. Click Analyze. Copy and paste the results to the text file. What is the sequence of the complementary strand? Compute it at http://arep.med.harvard.edu/labgc/adnan/projects/Utilities/revcomp.html. Note that the complementary sequence is also written from the 5' to 3' direction. Rewrite the complement from the 3' to 5' direction.

Part C: Obtaining the Plasmid Map from EMD Biosciences

1. Go to www.emdchemicals.com/life-science-research.
2. Search for pET-15b DNA. Click on the product to open. On the right, click to download the Vector map as a .pdf and open the TB045 pET-15b Vector map.

Part D: Obtaining the Cleavage Sequences for Restriction Enzymes That Can Be Used to Clone the Sequence to the Cloning Site from New England Biolabs

1. Go to www.neb.com/nebecomm/default.asp.
2. Search for the desired restriction enzymes, and record the cleavage sequences for each.

Part E: Using the Plasmid Map for pET-15b to Add the Restriction Enzyme Sites to the Appropriate Ends of the Foreign DNA to Be Cloned

1. Using the given plasmid map for pET-15b, select one or more restriction enzymes to clone the gene of interest into the multiple cloning site with *Nde*I, *Bam*HI, or *Xho*I after the promoter and His-tag. State which one(s) you chose. Add these sequence(s) from one of the restriction enzymes to the 5' or 3' end of the oligonucleotide as appropriate (upstream and downstream from the PowerPlex16 primer binding regions) to clone this gene in the orientation given by the original DNA sequence in the 5' to 3' direction.
2. Which enzyme is used to catalyze the formation of the covalent bonds between the DNA construct and the plasmid? (Hint: See Background.)

QUESTIONS

1. Define the following terms transcription, amplification, digestion, ligation, and transformation.
2. For each of these terms, state what enzyme, chemical, or treatment (if any) is used for the action.
3. State why you selected the restriction enzymes you did and why you added them to the 5' and 3' ends as you selected.

Reference

Frumpkin, D., Wasserstrom, A., Davidson, A., Grafit, A., 2010. Authentication of forensic DNA samples. Forensic Sci. Int. Genetics 4, 95–103.

21

Deoxyribonucleic Acid (DNA) Extraction from Botanical Material and Polymerase Chain Reaction (PCR) Amplification

OBJECTIVE

To learn how to extract DNA from botanical material from the *Cannabis sativa* (marijuana) plant and others and perform PCR to amplify a locus of interest to differentiate plant species.

SAFETY

Handle the SYBR Green reaction mix carefully; SYBR Green is an intercalating agent but is not noted to be carcinogenic. If you spill any chemicals on your person, wash them off immediately with soap and water.

The real-time PCR instrument is very sensitive to background noise. Do not touch the plate without gloved hands. To avoid contamination, wear gloves when handling the samples.

MATERIALS

1. Terra PCR Direct Polymerase Mix (Clontech) OR
2. DNeasy Plant Mini Kit (Qiagen) OR
3. Extract-N-Amp Kit (Sigma-Aldrich)
4. Proteinase K (20 mg/mL)
5. Gene-specific PCR primers: forward (5'- GAGGGTTTCTAATTTGTTATGTT-3'), reverse (5' -ACTAGAGGACTTGGACTATGTC-3')
6. 96-well PCR plates and microseal film or PCR tubes
7. Nuclease-free water
8. Extracted DNA template (1- to 10-ng range)

9. Various variable volume micropipettes (e.g., 0.5 to 10 μL, 10 to 100 μL, 100 to 1000 μL)
10. Autoclaved pipette tips for pipettes
11. Sterile microcentrifuge tubes or PCR tubes for preparing samples
12. Real-time PCR or traditional
13. Gloves, lab coats, and goggles
14. Hole punch
15. Forceps
16. Heating block or water bath
17. Agarose
18. 1X TAE buffer and DNA ladder
19. Gel boxes and power supply
20. CTAB
21. Silica

BACKGROUND

Cannabis sativa L. (Cannabaceae) is one of the earliest known cultivated plants. It is a schedule I drug in the United States but is accepted in 16 states and the District of Columbia for medicinal purposes. It is one of the highest dollar value crops grown in the United States and presents a significant caseload for forensic chemists tasked with quantifying and identifying the material.

The complete genomes for *Cannabis sativa* (400 million bases) and *Cannabis indica* were recently sequenced by a company, Medical Genomics, and published on Amazon's EC2 cloud computing service. An evaluation of single nucleotide polymorphism variations is being initiated and will likely be an advancing area of research. A set of microsatellite PCR primers has been published by Gilmore and Peakall in 2000 and Alghanim and Almirall in 2003.

In this laboratory, you will use a one-step extraction/PCR DNA amplification procedure or a two-step extraction/PCR procedure to evaluate a locus unique to the Cannabis plant. This laboratory experiment will employ one of many commercial kits available for this purpose. These kits are fast and provide good microgram DNA yields. Older methods employing silica, CTAB, and chloroform are well documented but require many more steps, toxic organics, and more time. However, the old method still produces the highest DNA yield at the least cost. The CTAB method will also clean cannabinoid compounds from the Cannabis DNA. PCR detection methods can be used to differentiate marijuana plants from other plant material that may be collected from the crime scene. They can also be used to trace the genetic origins of the plant material. The plant material may be fresh, frozen, or dried. The extracted DNA may be compared to other plants including spices purchased from a supermarket or standard biological specimens including Arabidopsis and other mustard plants (including cabbage, broccoli, cauliflower, turnip, rapeseed, mustard, radish, horseradish, cress, wasabi, and watercress).

Primers specific to *Cannabis sativa* (Figure 21.1) within the intergenic spacer between the *trn*L 3′ exon and *trn*F gene in the chloroplast only give a PCR product in the presence of *Cannabis sativa*. Cannabis can be detected by the successful production of the 197 bp PCR product (Linacre and Thorpe, 1998). Only the marijuana plant sample should yield the

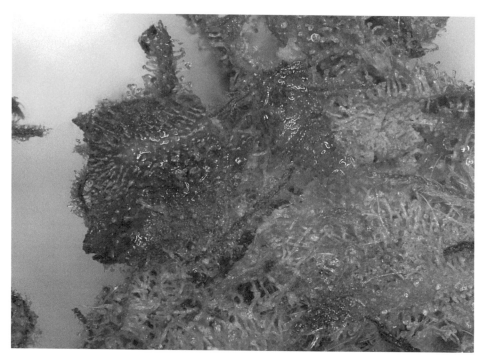

FIGURE 21.1   *Cannabis sativa* plant material (50x). *(Courtesy of Francesca Wheeler.)*

expected band produced from the PCR reaction on an agarose gel (as compared to Arabidopsis, Brassica, Zea, and other plants). Capillary electrophoresis can also be employed to analyze Cannabis samples (Coyle, 2012). Alternatively, primers in the AFLP Plant Mapping Kit (Applied Biosystems; Foster City, California) (Datwyler and Weiblen, 2006) and published STR markers can be used to identify marijuana (Howard et al., 2008, Gilmore et al., 2003).

PROCEDURE

Option 1 (Week 1)

Terra PCR Direct

1. Add a 1.2-mm diameter disc of plant tissue to a PCR tube or 96-well plate.
2. Pipet the following into the thin-walled PCR tube or 96-well plate for each sample:
   - 25 µL of 2x Terra PCR Direct Buffer (with Mg^{2+} and dNTPs)
   - 3 µL of 5 µM forward primer
   - 3 µL of 5 µM reverse primer
   - 18 µL of nuclease-free water
   - <u>1 µL of Terra PCR Direct Polymerase Mix</u>
   - 50 µL Total Reaction Volume

3. Prepare one negative (no template) control sample (substitute nuclease-free water for template DNA) (as a class).
4. Prepare one positive control sample (*Cannabis sativa* DNA, 1 ng in 1 μL for template DNA) (as a class).
5. Mix all tubes thoroughly by inverting or flicking tubes.
6. Centrifuge to remove bubbles. (If using 96-well plates, mixing may be performed using a sterile pipet tip, but care should be used to not introduce bubbles unless a plate centrifuge is available.)
7. Place all the tubes in the PCR Thermal Cycler, and begin the following preprogrammed cycles:

   98° C for 2 minutes (initial denaturation step is essential to denature Hot Start antibody)
   > 98° C for 10 seconds
   > 60° C for 15 seconds
   > 68° C for 60 seconds (60 sec. per kb)

   Repeat the preceding three-step cycle 30 to 40 times

   Hold at 4° C until the next morning. Your instructor will remove your samples and store them in the freezer for you until the next lab.

Option 2 (Week 1)

Extract-N-Amp Plant PCR Kit

1. Add a 0.5-cm diameter disc of plant tissue (punch with a regular office hole punch, clean before use and between samples with 70% ethanol) to a 2 mL microcentrifuge tube. If frozen plant tissue is used, keep samples on ice while punching discs.
2. Add 100 μL of the Extraction Buffer to the reaction tube, vortex to mix, and incubate at 95° C for 10 minutes. The leaf tissue may appear to be undegraded after this step.
3. Add 100 μL of the Dilution Solution; vortex to mix. (Store the sample in a refrigerator at 2° to 8° C, if desired.)
4. To a thin-walled PCR tube or 96-well plate for each sample, add the following:
   10 μL Extract-N-Amp PCR ReadyMix (with Mg^{2+} and dNTPs and JumpStart Taq DNA polymerase hot start Taq antibody)
   1.6 μL of 5 μM forward primer
   1.6 μL of 5 μM reverse primer
   12.8 μL of nuclease-free water
   <u>4 μL leaf disc DNA extract</u>
   20 μL Total Reaction Volume (add mineral oil if thermocycler does not have a heated lid)
5. Prepare one negative (no template) control sample (substitute nuclease-free water for template DNA) (as a class).
6. Prepare one positive control sample (Cannabis sativa DNA, 1 ng in 1 μL for template DNA) (as a class).
7. Mix all tubes thoroughly by inverting or flicking tubes.
8. Centrifuge to remove bubbles. (If using 96-well plates, mixing may be performed using a sterile pipette tip but care should be used to not introduce bubbles unless a plate centrifuge is available.)

9. Place all the tubes in the PCR Thermal Cycler, and begin the following preprogrammed cycles:

    94° C for 3 minutes (initial denaturation step is essential to denature Hot Start antibody)
    94° C for 60 seconds
       60° C for 60 seconds
       72° C for 60 seconds (60 sec. per kb)
Repeat the preceding three-step cycle 30 to 35 times.
    72° C for 12 minutes to finish elongation of all strands.
Hold at 4° C until the next morning. Your instructor will remove your samples and store them in the freezer for you until the next lab.

Option 3 (Week 1)

DNeasy Plant Mini Kit

1. Sample a 0.5-cm diameter disc of plant tissue (punch with a regular office hole punch, clean before use and between samples with 70% ethanol, $\leq$100-mg wet weight or $\leq$20-mg lyophilized tissue). If frozen plant tissue is used, keep samples on ice while punching discs.
2. Grind samples using a mortar and pestle.
3. Add 400 µL of AP1 buffer and 4 µL of RNase A to each sample. (Do not mix AP1 buffer and RNase A prior to use.)
4. Vortex samples and incubate for 10 minutes at 65° C. Invert the tubes occasionally (two to three times) during incubation.
5. Add 130 µl of AP2 buffer. Mix by inverting, and incubate for 5 minutes on ice.
6. Centrifuge the sample for 5 minutes at 20,000 x g (14,000 rpm) or, if not available, maximum centrifuge speed.
7. Pipette the lysed sample into the top of a QIAshredder spin column placed in a 2-ml microcentrifuge tube. Centrifuge for 2 minutes at 20,000 x g.
8. Transfer the eluent into a new tube, without disturbing the pellet if present. Add 1.5 volumes of AP3/E buffer (with ethanol added), and mix by pipetting. Transfer 650 µL of the mixture into a DNeasy Mini spin column, and elute in a 2 mL-microcentrifuge tube. Centrifuge for 1 minute at $\geq$6000 x g ($\geq$8000 rpm) or, if not available, maximum centrifuge speed. Discard the flow-through. Add the remaining sample to the DNeasy Mini spin column and spin again.
9. Transfer the spin column into a new 2-mL microcentrifuge tube. Add 500 µL of AW buffer, and centrifuge for 1 minute at $\geq$6000 x g. Discard the flow-through.
10. Add 500 µL of AW buffer (with ethanol added) a second time, and centrifuge for 2 minutes at 20,000 x g. Then, remove the spin column from the microcentrifuge tube carefully so that the column does become contaminated with the eluent.
11. Transfer the spin column to a new 1.5-mL or 2-mL microcentrifuge tube.
12. Add 100 µL of AE buffer for elution. Incubate for 5 minutes at room temperature (15° to 25° C) and centrifuge for 1 minute at $\geq$6000 x g. Repeat once. The eluent is the plant DNA-containing portion.
13. To a thin-walled PCR tube or 96-well plate for each sample, add the following:

12.5 µL of 2x iQ SYBR Green SuperMix (with Mg^{2+} and dNTPs)
1 µL of 5 µM forward primer
1 µL of 5 µM reverse primer
9.5 µL of nuclease-free water
<u>1 µL of plant DNA extract</u>
25 µL Total Reaction Volume (add mineral oil if thermocycler does not have a heated lid)

14. Prepare one negative (no template) control sample (substitute nuclease-free water for template DNA) (as a class).
15. Prepare one positive control sample (*Cannabis sativa* DNA, 1 ng in 1 µL for template DNA) (as a class).
16. Mix all tubes thoroughly by inverting or flicking tubes.
17. Centrifuge to remove bubbles. (If using 96-well plates, mixing may be performed using a sterile pipet tip, but care should be used to not introduce bubbles unless a plate centrifuge is available.)
18. Place all the tubes in the PCR Thermal Cycler, and begin the following preprogrammed cycles:

95° C for 3 minutes
  95° C for 15 seconds
  60° C for 60 seconds
  72° C for 60 seconds (60 seconds per kb)
Repeat the above three-step cycle 30 to 40 times.

Hold at 4° C until the next morning. Your instructor will remove your samples and store them in the freezer for you until the next lab.

Option 4 (Week 1)

Organic-CTAB Extraction Protocol for Plant Material

1. Add 1200 µL 2x CTAB reagent with 1% ethanol by volume and 2 µL RNase stock and mix.
2. To each frozen (with liquid nitrogen) 0.1- to 0.25-g sample, grind with a mortar and pestle, and add to a 1.5-mL microcentrifuge tube.
3. Grind each sample separately with a blue micropestle, leaving the pestle in the tube.
4. Mix into the sample tube.
5. Place in 37° C incubation for 5 minutes to overnight.
6. Add 500 µL of chloroform and mix for at least 1 minute. Solution should be cloudy. Centrifuge for 1 to 5 minutes maximum speed.
7. Transfer upper, aqueous, phase into a new sterile 1.5-mL tube. Do not transfer any solid material to the new tube.
8. Add 200 to 800 uL of silica bind mix, and mix thoroughly for 5 minutes. Centrifuge briefly.
9. Discard the supernatant into beaker.
10. Let stand for a minute. Centrifuge. Discard liquid.
11. Lay tubes on sides at 37° C for 10 minutes.

12. Add 0.1 vol of 5 M NaCl; then, to the total sample volume, add 0.666 vol of 2-propanol. Mix at least 1 minute, microcentrifuge at least 1 minute, pour off the supernatant, and rinse the pellet with 1 mL of 70% ethanol.
13. Microcentrifuge again, discard the supernatant, and dry the pellet.
14. Resuspend the DNA pellet in 20 to 50 μL TE.
    If doing PCR same day, leave tubes at room temperature for 1 hour.
    Otherwise, store DNA solutions at 4° C.

Agarose Gel Electrophoresis (Week 2)

1. Prepare a 1% agarose gel. Combine 0.5 g of agarose and 50 mL of 1x TAE buffer (better resolution of DNA fragments than TBE buffer) in an Erlenmeyer flask covered with plastic wrap that has been punctured with a small hole.
2. Heat in the microwave for 40 seconds or until it boils and agarose is fully dissolved.
3. Remove from microwave.
4. Wait for it to cool to the touch, then pour it into the gel box positioned so that a rectangular box with four sides is created. Add a 14-well comb.
5. Once the gel is set (translucent blue color), rotate the gel so that the wells are positioned nearest the negative pole (black).
6. Fill the gel box with 1x TAE buffer to cover the gel.
7. Remove the comb from the gel.
8. To prepare samples for the gel, add 15 μL of the 6x loading buffer (if whole plant tissue was used, add 5 μL of 20 mg/mL Proteinase K to the 6X gel loading buffer to digest proteins in the cell debris that can encapsulate DNA product) to each 50-μL sample in the PCR tubes or 96-well plate.
9. Load 11 μL of the amplified samples-loading buffer to separate wells carefully by pipetting. Do not release the plunger until the pipette is removed from the gel so as to not draw up the samples.
10. To prepare the ladder for the gel, combine 8 μL of the DNA Logic ladder with 1 μL of 100x SYBR Green I dye and combine 4 μL of the Lonza 20 bp ladder with 1 μL of 100x SYBR Green I dye and 1 μL of the 6X gel loading buffer.
11. Load the entire portion of the two ladders to two different lanes on opposite sides of the gel.
12. Place the safety cover on the gel box to run the sample toward the positive (red) electrode. Connect and turn on the power supply.
13. Run the gel for 30 to 60 minutes at no more than 1 to 5 V/cm (or 150 V for 10-cm gel).
14. Turn off the power supply and unplug it. Remove the gel safety cover.
15. While wearing gloves, remove the gel from the casting plate, place it on a UV light box, and take a picture using a digital photodocumentation system with a SYBR filter.

QUESTIONS

1. Construct a graph of log base pairs vs. distance traveled for DNA ladder and add linear regression line.

2. Use linear regression line equation and distance traveled by unknown to determine amplicon size and presence or absence of marijuana for all samples.
3. Identify sample by lane which contained marijuana.

References

Alghanim, H.J., Almirall, J.R., 2003. Development of microsatellite markers in *Cannabis sativa* for DNA typing and genetic relatedness analyses. Anal. Bioanal. Chem. 376, 1225–1233.

Coyle, H.M., 2012. Capillary electrophoresis of DNA from *Cannabis sativa* for correlation of samples to geographic origin. Method Mol. Biol. 830, 241–251.

Datwyler, S.L., Weiblen, G.D., 2006. Genetic Variation in Hemp and Marijuana (*Cannabis sativa* L.) According to Amplified Fragment Length Polymorphisms. J. Forensic Sci. 51, 371–375.

Gilmore, S., Peakall, R., 2000. Isolation of microsatellite markers in *Cannabis sativa* L. (marijuana). Mol. Ecol. Notes 3, 105–107.

Gilmore, S., Peakall, R., Robertson, J., 2003. Short tandem repeat (STR) DNA markers are hypervariable and informative in *Cannabis sativa*: implications for forensic investigations. Forensic Sci. Int. 131, 65–74.

Howard, C., Gilmore, S., Robertson, J., Peakall, R., 2008. Developmental Validation of a *Cannabis sativa* STR Multiplex System for Forensic Analysis. J. Forensic Sci. 53, 1061–1067.
http://onlinelibrary.wiley.com/doi/10.1111/j.1556-4029.2008.00792.x/full?url_ver=Z39.88-2004&rft_val_fmt=info%3Aofi%2Ffmt%3Akev%3Amtx%3Ajournal&rft.genre=article&rft.jtitle=Mol%20Ecol%20Notes&rft.atitle=Isolation%20of%20microsatellite%20markers%20in%20Cannabis%20sativa%20L.%20%28marijuana%29&rft.volume=3&rft.issue=1&rft.spage=105&rft.epage=7&rft.date=2003&rft.aulast=Gilmore&rft.aufirst=S&rfr_id=info%3Asid%2Fwiley.com%3AOnlineLibrary

Linacre, A., Thorpe, J., 1998. Detection and identification of cannabis by DNA. Forensic Sci. Int. 91, 71–76.

CHAPTER

22

Social, Ethical, and Regulatory Concerns

When deciding to use deoxyribonucleic acid (DNA) evidence, the crime scene investigator and the forensic lab analyst must balance two alternative "goods." On one side, DNA can identify criminal perpetrators. On the other side, many courts and most citizens believe in a firm right to privacy.

Repeat offenders commit approximately 40% of all crimes. The apprehension of criminals serves society by preventing further crimes by those individuals and in turn reduces the number of victims. Countries and states all differ in how they respond to these issues.

There are three major forensic DNA databases of individuals convicted of a crime: the Combined DNA Indexing System (CODIS), which is maintained by the U.S. Federal Bureau of Investigation (FBI); the European Network of Forensic Science Institutes (ENFSI) DNA database; and the Interpol Standard Set of Loci (ISSOL) database maintained by Interpol.

Responses to the expansion of the CODIS database and the United Kingdom National DNA Database (NDNAD) range from excitement about DNA's potential and fear regarding its use. As the databases have been expanded and their uses extended, concerns about the ethical and social implications of those decisions have amplified.

These alternative goods raise a number of important issues:

- Can DNA samples be taken from individuals without prior consent?
- Should DNA samples be required from individuals convicted of crimes?
- Should the collection of DNA samples be restricted to individuals convicted of sex crimes, aggravated assault, violent crimes, and homicides, or should they be extended to individuals convicted of burglaries, nonsexual assaults, drug crimes, felonies, and misdemeanors?
- Should DNA samples be required from individuals suspected of and arrested for sex crimes, aggravated assault, violent crimes, homicides, burglaries, nonsexual assaults, drug crimes, felonies, and misdemeanors?
- Should DNA samples be disposed of if a suspect/arrestee is exonerated from a crime, or should those samples be retained? Should the data be expunged from databases?
- What should be the time interval between exoneration of an individual and expunging of the related database entry?

Forensic DNA Biology

181

- Should DNA databases separate entries of individuals convicted of crimes and those suspected of crimes? Should it be organized by type of crime?
- Should databases of missing persons and their family members be separated from offender databases?
- Should DNA samples be collected from minors who have committed crimes?
- Should DNA samples be collected from all members of the population for the purpose of understanding the individuality of a profile within a database?
- Does profiling the entire population turn the entire population into suspects?
- Is the evaluation of DNA evidence essential for solving crimes?
- If there is a gene or set of genes that predisposes individuals to a life of crime, should the population as a whole be evaluated for possession of the gene(s)?
- Should persons possessing a gene or set of genes predisposing the individual to criminal acts be targeted first in solving crimes?
- Should collected DNA be evaluated for medical or character disorders of the individual for use in solving crimes?
- Should forensic DNA phenotyping (a method in which crime scene DNA is analyzed to compose a description of the unknown suspect, including external and behavioral features, geographic origin, and perhaps surname) be permitted?
- Is the collection and use of DNA a modest invasion of privacy but one that is acceptable for the common good?
- Could some of these procedures be covered by "objectivity of science" to protect their use?

There are some central differences between the collection and use of DNA for medical purposes and its collection and analysis for forensic purposes. First, DNA used for medical purposes is often donated, whereas DNA for forensic purposes is often taken from suspects against their will. For medical uses, anonymity is guaranteed and legally required, whereas the use of DNA to solve a crime is associable with personal identity.

There are three distinct points of view regarding the collection, retention, and possible purging of genetic evidence. These include genetic exceptionalism, genetic minimalism, and biological pragmatism.

Genetic exceptionalists view genetic samples as "personal" material and believe that the genetic information contained therein is "private." Exceptionalists are uncomfortable with the retention of genetic profiles in databases and are even more uncomfortable with the storage of bodily fluids. The important question for this group is, does the biological sample constitute personal data or does the information derived from the biological sample constitute personal data?

Genetic minimalists view DNA profiles as a set of numbers—no more damaging than profiling based on height and weight. These numbers, of course, represent numbers of copies of short tandem repeats analyzed by DNA analysts. The information capacity of these probed sites is minimal, and a match or exclusion of these numbers can be used to solve crimes. The DNA profile is seen as no more informative than a traditional fingerprint or barcode or even a photograph. Here, the DNA profile is seen as one of many individualizing features.

Biological pragmatists consider DNA and biological fluids and materials left at crime scenes to be abandoned property and collecting such evidence a routine and mundane part of a crime scene investigation and the work of crime scene investigators through legal search and seizure protocols. Laboratory analysts are tasked with evaluating the evidence to best solve the mystery of the crime. The approach to each crime may be individualized to solve the crime and minimize the use of lab resources and time. The minimalists often ignore current research to extend the ability of DNA to individualize persons using other loci for eye color, hair color, inherited disease, and other traits, for example.

The point of view of the individual will greatly influence responses to the questions listed previously and will impact social and political attitudes to criminalistics. Depending on the level of protection viewed as appropriate for a "victim," "innocent," or "criminal" individual, attitudes may change over time. The power of a large DNA databank and stored samples extends to the possibility of it acting as a deterrent for a "criminal" or an "innocent" individual who could one day be a "criminal." A match of DNA evidence from one crime scene to another crime scene found through the power of the database could assist law enforcement officials and save taxpayers time, money, and other resources. As the Innocence Project has demonstrated, DNA evidence can also exonerate individuals falsely accused of crimes and prove their innocence. The innocent individuals will suffer from publicity related to the arrest and search of their person and background, but the question remains as to whether the violation of their human rights by DNA testing is different than that encountered by fingerprinting or conventional police work.

Today, forensic databases are genomically minimalist. That is not to say that computational power will not allow more information to be stored. Genomic minimalism can raise questions in cases where two individuals cannot be resolved using the current multiplex loci, especially if the original biological material is not retained. In Arizona, one pair of individuals was found to match at 12 loci. In Maryland, three pairs were found to match at all 13 CODIS loci. Hundreds of pairs have been found to match at 9 CODIS loci in state databases including Illinois (Geddes 2010). The original biological material would need to be probed to differentiate the two persons. Retained samples can be used in reprofiling and error checking. They can also be employed in quality assurance and training programs.

The potential advantages of storing arrestee DNA are significant. For example, crimes are often committed by individuals who have a prior criminal record. Innocent people could be tested using DNA profiling prior to incarceration for crimes they did not commit. However, there are also disadvantages to storing DNA of arrested individuals. Arrestees may be exonerated and thus some believe that they should have a process for purging their DNA profile from the database. Partial or even complete matches could implicate an innocent person in a crime, even a prior criminal who was not involved in the current crime. Genetic information may be extended to phenotypic and medical profiling that could be used for unintended reasons (e.g., denial of health insurance) or for the determination of family relationships without the family members' consent. Ethnic minorities are often overrepresented in DNA criminal databases; this could lead to ethnic bias in the criminal justice system. Even the most secure database has a chance of being compromised.

The following are U.S. federal laws that govern DNA evidence (Table 22.1). State laws can be found at www.dna.gov/statutes-caselaw/state-statutes.

In Europe, Article 8 of the European Convention on Human Rights defines guidelines for the protection of individual rights. The Prüm decision by the European Commission obliged all countries in the European Union to make their forensic DNA databases searchable for use by authorities in other EU countries (on a hit/no hit basis) (Prainsack & Toom 2010).

TABLE 22.1   U.S. Federal Laws that Govern DNA Evidence

Law	Date	Implication
DNA Fingerprint Act	2005	Authorized collection of DNA samples from persons arrested or detained under Federal authority, established an opt-out system for the expungement of DNA from CODIS
Justice For All Act	2004	Established enforceable victims' rights, authorized grants to reduce the DNA backlog, enhanced DNA collection and analysis efforts, provided for postconviction DNA testing for those who maintain their innocence, and authorized grants to improve the quality of representation in state capital cases
DNA Backlog Elimination Act (42 U.S.C. §14135a)	2000	To make grants to States for carrying out DNA analyses for use in the Combined DNA Index System of the Federal Bureau of Investigation, to provide for the collection and analysis of DNA samples from certain violent and sexual offenders for use in such system
Crime Information Technology Act	1996	To provide for the improvement of interstate criminal justice identification, information, communications, forensics, and provide for DNA identification grant programs
DNA Identification Act (42 U.S.C. §14132)	1994	DNA data is confidential, authorized NDIS

References

Geddes, B., 2010. Unreliable evidence? Time to open up DNA databases. New Scientist 2742, http://www.newscientist.com/article/mg20527424.700-unreliable-evidence-time-to-open-up-dna-databases.html?full=true

Prainsack, B., Toom, V., 2010. The Prüm regime: Situated dis/empowerment in transnational DNA profile exchange. British Journal of Criminology 50, 1117–1135.

Selected Forensic DNA Biology Case Studies

HISTORICAL DNA ANALYSIS

1. Fisher, D.L., Holland, M.M., Mitchell, L., Sledzik, P.S., Webb Wilcox, A., Wadhams, M., Weedn, V.W., 1993. Extraction, Evaluation, and Amplification of DNA from Decalcified and Undecalcilfied United States Civil War Bone. Journal of Forensic Sciences 38, 60–68.
2. Ivanov, P.L., Wadhams, M.J., Roby, R.K., Holland, M.M., Weedn, V.W., Parsons, T.J., 1996. Mitochondrial DNA sequence heteroplasmy in the Grand Duke of Russia Georgij Romanov establishes the authenticity of the remains of Tsar Nicholas II. Nature 12, 417–420.
3. Foster, E.A., Jobling, M.A., Taylor, P.G., Donnelly, P., de Knijff, P., Mieremet, R., Zerjal, T., Tyler-Smith, C., 1998. Jefferson fathered slave's last child. Nature 396, 27–28.
4. Jehaes, E., Toprak, K., Vanderheyden, N., Pfeiffer, Cassiman, J.-J., Brinkmann, B., Decorte, R., 2001. Pitfalls in the analysis of mitochondrial DNA from ancient specimens and the consequences for forensic DNA analysis: the historical case of the putative heart of Louis XVII. Int. J. Legal Med. 115, 135–141.
5. Wandeler., P., Smith, S., Morin, A., Pettifor, R.A., Funk, S.M., 2003. Patterns of nuclear DNA degeneration over time- a case study in historic teeth samples. Molecular Ecology 12, 1087–1093.
6. Parson, W., Brandstatter, A., Niederstatter, H., Grubwieser, P., Scheithauer, R., 2007. Unravelling the mystery of Nanga Parbat. Int. J. Legal Med. 121, 309–310.
7. Dissing, J., Binladen, J., Hansen, A., Sejrsen, B., Willerslev, E., Lynnerup, N., 2007. The last Viking King: A royal maternity case solved by ancient DNA analysis. Forensic Science International 166, 21–27.
8. Milde-Kellers, A., Krawczak, M., Augustin, C., Boomgaarden-Brandes, K., Simeoni, E., Kaatsch, H.-J., Mühlbauer, B., Schuchardt, 2008. An Illicit Love Affair During the Third Reich: Who is My Grandfather? J. Forensic Sci. 53, 377–379.
9. Coble, M.D., Loreille, O.M., Wadhams, M.J., Edson, S.M., Maynard, K., Meyer, C.E., Niederstatter, H., Berger, C., Falsetti, A.B., Gill, P., Parson, W., Finelli, L.N., 2009. Mystery Solved: The Identification of the Two Missing Romanov Children Using DNA Analysis. PloS One 4, e4838.
10. Loreille, O.M., Parr, R.L., McGregor, K.A., Fitzpatrick, C.M., Lyon, C., Yang, D.Y., Speller, C.F., Grimm., M.R., Grimm, M.J., Irwin, J.A., Robinson, E.M., 2010. Integrated DNA and Fingerprint Analyses in the Identification of 60-Year-Old Mummified Human Remains Discovered in an Alaskan Glacier. In: J. Forensic Sci. 55 813–818.

PATERNITY TESTING

11. Junge, A., Brinkmann, B., Fimmers, R., Madea, B., 2006. Mutations or exclusion: an unusual case in paternity testing. Int. J. Legal Med. 120, 360–363.
12. Lukka, M., Tasa, G., Ellonen, P., Moilanen, K., Vassiljev, V., Ulmanen, I., 2006. Triallelic patterns in STR loci used for paternity analysis: Evidence for a duplication in chromosome 2 containing the TPOX STR locus. Forensic Science International 164, 3–9.
13. Narkuti, V., Oraganti, N.M., Vellanki, R.N., Mangamoon, L.N., 2009. De novo deletion at D13S317 locus: A case of paternal–child allele mismatch identified by microsatellite typing. Clinica Chimica Acta 403, 264–265.
14. González-Andrade, F., Sánchez, D., Penacino, G., Jaretta, B.M., 2009. Two fathers for the same child: A deficient paternity case of false inclusion with autosomic STRs. Forensic Science International: Genetics 3, 138–140.

DNA IN MODERN FORENSIC CASES

15. Tsutsumi, H., Katsumata, Y., 1993. Forensic study of stains of blood and saliva in a chimpanzee bite case. Forensic Science International 61, 101–110.
16. Menotti-Raymond, M., David, V.A., O'Brien, S.J., 1997. Pet cat hair implicates murder suspect. Nature 386, 774.
17. Sweet, D., Hildebrand, D., 1999. Saliva from cheese bite yields DNA profile of burglar: a case report. Int. J. Legal Med. 112, 201–203.
18. Oikawa, H., Tun, Z., Young, D.R., Ozawa, H., Yamazaki, K., Tanaka, E., Honda, K., 2002. The specific mitochondrial DNA polymorphism found in Klinefelter's syndrome. Biochemical and Biophysical Research Communications 297, 341–345.
19. Steinlechner, M., Berger, B., Niederstatter, H., Parson, W., 2002. Rare failures in the amelogenin sex test. Int. J. Legal Med. 116, 117–120.
20. Góes, A.C., de, S., da Silva, D.A., Domingues, C.S., Sobrinho, J.M., de Carvalho, E.F., 2002. Identification of a criminal by DNA typing in a rape case in Rio de Janeiro, Brazil. Sao Paulo Med. J. 120, 77–80.
21. Schulz, M.M., Wehner, H.-D., Reichert, W., Graw, M., 2004. Ninhydrin-dyed latent fingerprints as a DNA source in a murder case. Journal of Clinical Forensic Medicine 11, 202–204.
22. Grubwieser, P., Pavlic, M., Gunther, M., Rabl, W., 2004. Airbag contact in traffic accidents: DNA detection to determine the driver identity. Int. J. Legal Med. 118, 9–13.
23. Drexler, C., Glock, B., Vadon, M., Staudacher, E., Dauber, E.-M., Ulrich, S., Reisacher, R.B.K., Mayr, W.R., Lanzer, G., Wagner, T., 2005. Tetragametic chimerism detected in a healthy woman with mixed field agglutination reactions in ABO blood grouping. Transfusion 45, 698–703.
24. Negi, D.S., Alam, M., Bhavani, S.A., Nagaraju, J., 2006. Multistep microsatellite mutation in the maternally transmitted locus D13S317: a case of maternal allele mismatch in the child. Int. J. Legal Med. 120, 286–292.
25. Narkuti, V., Vellanki, R.N., Anubrolu, N., Doddapaneni, K.K., Kaza, P.C.G., Mangamoori, L.N., 2008. Single and double incompatibility at vWA and D8S1179/D21S11 loci between mother and child: Implications in kinship analysis. Clinica Chimica Acta 395, 162–165.
26. Castella, V., del Mar Lesta, M., Mangin, P., 2009. One person with two DNA profiles: a(nother) case of mosaicism or chimerism. Int. J. Legal Med. 123, 427–430.
27. Nurit, B., Anat, G., Michal, S., Lilach, F., Maya, F., 2011. Evaluating the prevalence of DNA mixtures found in fingernail samples from victims and suspects in homicide cases. Forensic Science International: Genetics 5, 532–537.

MASS DISASTER CASES

28. Budimlija, Z.M., Prinz, M.K., Zelson-Mundorff, A., Wiersema, J., Bartelink, E., MacKinnon, G., Nazzaruolo, B.L., Estacio, S.M., Hennessey, M.J., Shaler, R.C., 2003. World Trade Center Human Identification Project: Experiences with Individual Body Identification Cases. Croatian Medical Journal 44, 259–263.
29. Dzijan, S., Primorac, D., Marcikic, M., Andelinovic, S., Sutlovic, S., Dabelic, S., Lauc, G., 2005. High Estimated Likelihood Ratio Might Be Insufficient in a DNA-lead Process of Identification of War Victims. Croatica Chemica Acta 78, 393–396.
30. Calacal, G.C., Delfin, F.C., Tan, M.M.M., Roewer, L., Magtanong, D.L., Lara, M.C., Fortun, R.dR., De Undria, M.C.A., 2005. Identification of Exhumed Remains of Fire Tragedy Victims Using Conventional Methods and Autosomal/Y-Chromosomal Short Tandem Repeat DNA Profiling. The American Journal of Forensic Medicine and Pathology 26, 285–291.

DNA MIXTURES

31. Liao, X.H., Lau, T.S., Ngan, K.F.N., Wang, J., 2002. Deduction of paternity index from DNA mixture. Forensic Science International 128, 105–107.
32. Hatsch, D., Amory, S., Keyser, C., Hienne, R., Bertrand, L., 2007. A Rape Case Solved by Mitochondrial DNA Mixture Analysis. J. Forensic Sci. 52, 891–894.

MTDNA ANALYSIS

33. Cano, R.J., Poinar, H.N., Pieniazek, N.J., Acra, A., Poinar Jr., G.O., 1993. Amplification and sequencing of DNA from a 120-135-million-year-old weevil. Nature 363, 536–538.

34. Parson, W., Pegoraro, K., Niederstatter, H., Foger, M., Steinlechner, M., 2000. Species identification by means of the cytochrome b gene. Int. J. Legal Med. 114, 23–28.

MICROBIAL DNA ANALYSIS

35. Higgins, J.A., Cooper, M., Schroeder-Tucker, L., Black, S., Miller, D., Karns, J.S., Manthey, E., Breeze, R., Perdue, M.L., 2003. A Field Investigation of *Bacillus anthracis* Contamination of U.S. Department of Agriculture and Other Washington, D.C., Buildings during the Anthrax Attack of October 2001. Applied and Environmental Microbiology 69, 593–599.
36. Beecher, D.J., 2006. Forensic Application of Microbiological Culture Analysis To Identify Mail Intentionally Contaminated with *Bacillus anthracis* Spores. Applied and Environmental Microbiology 72, 5304–5310.

FORENSIC BOTANY

37. Virtanen, V., Korpelainen, H., Kostamo, K., 2007. Forensic botany: Usability of bryophyte material in forensic studies. Forensic Science International 172, 161–163.

WILDLIFE FORENSICS

38. Lorenzini, R., 2005. DNA forensics and the poaching of wildlife in Italy: A case study. Forensic Science International 153, 218–221.

MEDICAL FORENSICS

39. Dong-ling, T., Yan, L., Xin, Z., Xia, L., Fang, Z., 2009. Multiplex fluorescent PCR for noninvasive prenatal detection of fetal-derived paternally inherited diseases using circulatory fetal DNA in maternal plasma. European Journal of Obstetrics and Gynecology and Reproductive Biology 144, 35–39.

TOUCH DNA

40. Schwark, T., Poetsch, M., Preusse-Prange, A., Kamphausen, T., von Wormb-Schwark, N., 2012. Phantoms in the mortuary—DNA transfer during autopsies. Forensic Science International 216, 121–126.

Index

Page numbers followed by "f" indicate figures and "t" indicate tables.